The Moving Body

The Moving Body

PAT KENNEDY

M.C.S.P., Dip.T.P., Dip.Phys.Ed.

With a Foreword by

MONNICA C. STEWART

M.B., B.S., D.(Obst.) R.C.O.G.
Assistant Physician, Department of Geriatric Medicine,
Edgware General Hospital

FABER AND FABER

London Boston

First published in 1979
by Faber and Faber Limited
3 Queen Square London WC1N 3AU
Printed in Great Britain by
Cox & Wyman Limited,
London, Fakenham and Reading

British Library Cataloguing in Publication Data

Kennedy, Pat
The moving body.
1. Human physiology
I. Title
612 QP34.5

ISBN 0–571–11346–X

Contents

Illustrations

Illustrations

Illustrations

Foreword

It is with a great sense of pleasure, privilege and awe, that I write a foreword to this splendid little book.

Pleasure because, in recent years, I have become interested and involved in different forms of movement classes and groups. During this time I have met many fine teachers of movement, who have been avid to learn more about basic anatomy, physiology, body mechanics and the explanation of some of the commoner physical problems that they encounter in their classes. This book, so simple, clear and straightforward, will fill a long felt want for them. It certainly would have helped me if it had been available at the beginning of my own medical student career. Pat Kennedy has extracted the essentials needed to understand body movement, breathing and circulation and linked them together with clear diagrams. She has also very succinctly described some of the disabilities that class leaders may meet amongst their members and this additional knowledge will help to make non-medically trained people feel more comfortable in their teaching approach.

Anyone, regardless of their previous education, who reads this book through should by the end have a very clear idea of how their body moves.

My sense of privilege comes in being asked to write this foreword because I myself am a tyro in the movement field. My main professional interest has lain in endeavouring to improve the quality of life for disabled and elderly people and it is only of recent years that I have come to realise the vast contribution that movement classes and groups can make in this direction. To see a group of heavily handicapped elderly people (particularly ones who may have been living for some time in an institution) enjoying and visibly expanding their physical horizons in a dynamic music and movement class is a very happy and warming experience.

It is with awe that I congratulate the author on refining so clearly and simply such a huge subject to the basic essentials of anatomy and physiology needed to help understanding. It is a remarkable feat for which many movement teachers will be grateful and from which many thousands of class members will benefit in the future.

MONNICA C. STEWART
1979

Preface

This small book has been written to help leaders of movement classes and groups, to have a simple understanding of how the body works and how it reacts to movement. The word 'leader' is used throughout to mean the person who conducts the class – this could well include nurses, nursing auxiliaries, physiotherapists, physiotherapy aides, occupational therapists as well as Keep Fit leaders, teachers of modern dance and many others.

The content of the book is deliberately basic and I hope that it will help all leaders to be more confident in their approach to classes which have disabled or elderly people included. Movement classes for the elderly are increasing in popularity and there is now considerable interest in this field of recreational movement among local authorities, National Associations as well as in the long-stay hospitals.

I am particularly grateful to Kay Evans M.B.E., Project Officer, Physical Recreation and Sport, the Disabled Living Foundation, for cheerfully reading proofs and for generous help whenever needed, and to Joan P. Adams, Principal of the Marylebone Institute, for her enthusiasm and support in this work. Dr. Monnica Stewart, Assistant Physician to the Department of Geriatric Medicine, Edgware General Hospital, has been the driving force behind the publishing of this volume; for this and for kindly writing the Foreword, I express my sincere thanks.

PAT KENNEDY
1979

CHAPTER 1

Tissues Forming the Body

The body is made up of a skeletal (bony) structure together with various types of soft tissues, such as skin, fat and muscle. Some tissues are highly specialised in function and the main characteristics of the relevant tissues are described.

BONE

Bone forms the hard framework of the body. As can be seen in any anatomy book, or felt in the living body, bones vary in shape and size, and fulfil a variety of functions. Bone depends for its hardness on inorganic salts, e.g. calcium. Bone is not necessarily dense, but is always more dense at areas of the body where extra strength is required, e.g. the leg bones. The bones of the arms are less dense and lighter, as they are used in a more mobile manner, in reaching, dressing and feeding. Bone is nourished by a blood supply, and under normal conditions is fairly resistant to breakage by everyday stresses.

The main features of typical bones and the terminology used are shown in Figure 1/1.

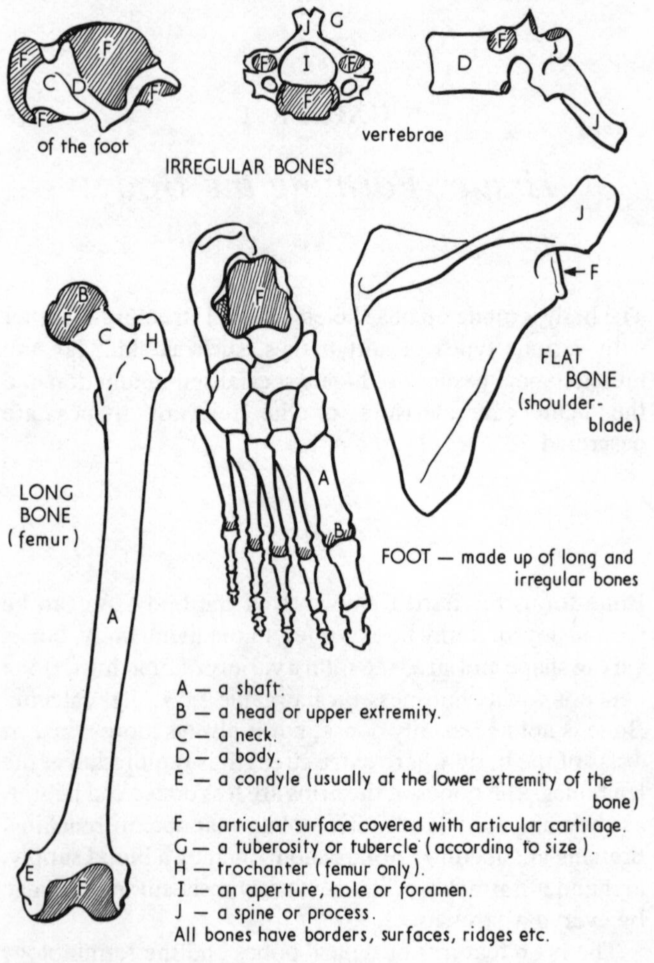

Fig. 1/1 Diagrams showing the main types of bones of the body

CARTILAGE

Cartilage is a dense hard tissue comparable in its toughness with bone. It is, however, flexible and resilient. There are two types of cartilage.

Hyaline cartilage–articular (*see* Fig. 1/1)

This type of cartilage is bluish-white and glistening; it covers the articular surfaces of bones. It is hard wearing and maintains a smooth surface during movement of one bone on another. With advancing years this articular cartilage tends to wear in varying degrees and the joint then becomes less easy to move; this is most likely in joints which have to bear weight, e.g. leg or lower back. Fluid in the joint as well as a blood supply nourishes articular cartilage.

Fibro-cartilage

This type is more fibrous in texture; it is tough, inelastic and resistant to stretching; when used as a pad it is compressible, like a fairly firm rubber. This is important to remember as the discs between adjacent vertebra are made of fibro-cartilage. The function and importance of these discs in the movement of the spine will be discussed later.

LIGAMENT

Ligamentous tissue is strong, fibrous and pliable. Ligaments are associated with joints; the bone ends are held loosely together by means of ligaments. They are extra strong where stress and strain are most likely to occur, so protecting the joint from damage. The two bone ends of a joint are enveloped in a ligamentous sleeve, called the capsule. The capsule is reinforced by extra bands of fibrous tissue for further protection; e.g. the knee joint has reinforcing ligaments on the inside and outside of the knee to prevent any sideways movement or twisting strains.

Ligaments are not intended to bind bones close together, but to allow movement within a limited range. The main security of a joint is the muscles which pass over it and support it. A few ligaments in the body contain some elastic fibres and these ligaments are slightly elastic in function, e.g. the 'spring ligament' on the inner side of the foot which helps to support the inner longitudinal arch of the foot.

CONNECTIVE TISSUES

There are various forms of connective tissue, such as fat, elastic tissue, and fibrous tissue. Their role is mainly one of support for specialised structures, forming sheaths for muscles, and dividing muscle groups. Connective tissues are found in almost every part of the body.

MUSCLE

There are three types of muscle tissue in the body:

1 *Cardiac muscle* from which the heart is formed.

2 *Involuntary muscle* which is found in blood vessels, digestive organs, and the respiratory tract. As the name implies, the activity of involuntary muscle is not under the will or control of a person.

3 *Voluntary or skeletal muscle*. This is muscle which is responsible for willed, voluntary movement and is under

the control of the individual. This type is of major interest for those concerned with any form of activity.

A single voluntary muscle is composed of many muscle fibres lying parallel and in bundles. These tiny fibres or cells are like strands of cotton, some thicker than others and varying in length. The muscles formed are different in shape, size and contain a varying number of muscle fibres (*see* Fig. 1/2).

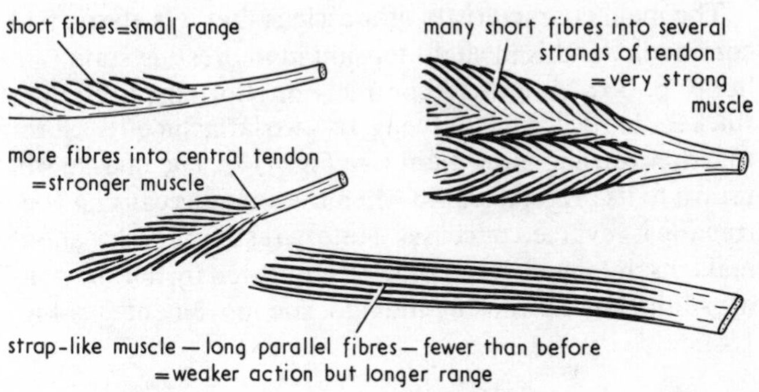

Fig. 1/2 Diagram showing the different formations of muscles

A muscle is seen to consist of bundles of muscle fibres giving bulk and shape, collected together and enveloped in a sheath of fibrous tissue. Muscles are attached at either end to one of the following structures:

a) to bone by a fibrous tendon

b) to bone, by its fibres directly to the bone covering – the periosteum; or by fibrous tissue

c) to fibrous tissue of other muscles, e.g. the abdominal muscles (*see* Fig. 1/3)

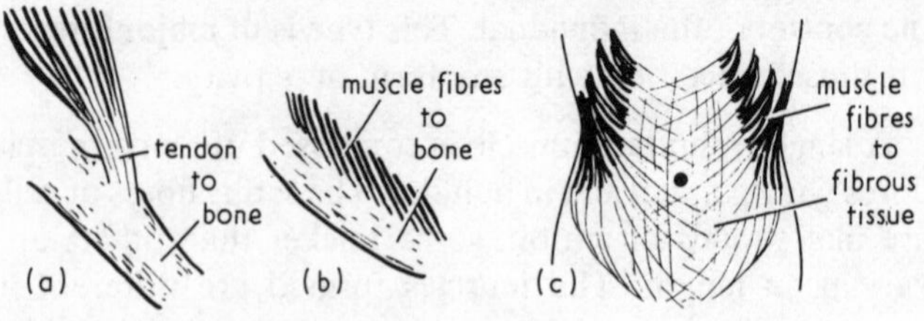

Fig. 1/3 Diagram showing the different types of muscle attachments

The main characteristic of muscle is that it is elastic and contractile, and is in slight tension in the resting state (*see* 'tone' p. 53). A muscle contracts or shortens, becoming thicker, harder, and drawing its two attachments closer together causing movement (*see* Fig. 1/4). The muscle will return to its original length when the stimulus causing contraction is reversed or ceases. Note: a resting muscle cannot make itself longer, it can only be elongated by the shortening of the muscle causing the opposite movement (*see* Fig. 1/4).

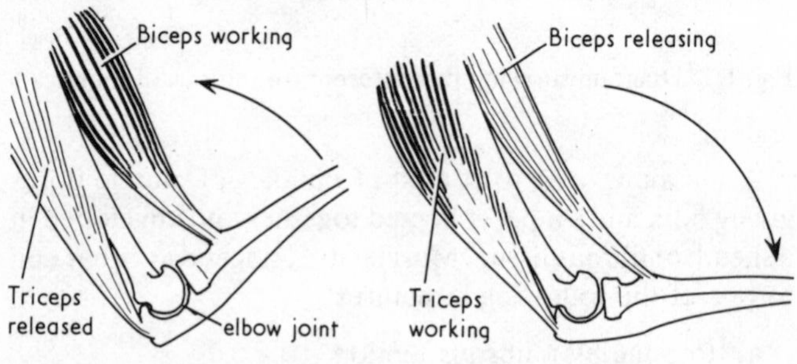

Fig. 1/4 Diagram to show the mechanism of muscle action

Muscle fibres respond to the 'All or None' law, i.e. each fibre is capable only of full contraction or no contraction. A muscle fibre cannot contract slightly. To lift a weight a few muscle fibres in the appropriate muscle will contract maximally; the rest will remain inactive. To lift a heavier weight more fibres will be brought into action, until finally the heaviest weight possible will demand maximal contraction of all the fibres in that muscle. This is the work limit of which that muscle is capable. This can be verified practically using the biceps muscle; a) bend the elbow, without any weight in the hand; b) repeat the bend holding a one pound (1 lb) weight; c) repeat, this time holding a ten pound (10 lb) weight; notice the shape, size and hardness of the muscle at each lift. From this example it is evident that the response of a muscle is related to the demand made upon it. It is important to place sufficient demand on muscles in order to maintain their efficiency. As age advances muscle power deteriorates, therefore the demand must be moderated to meet this.

Specialised tissues, with the exception of bone, do not repair themselves with special tissue but with a type of non-specialised fibrous tissue or scar tissue; a torn muscle healing by a scar will be less powerful in its action.

The whole mechanism of voluntary movement is controlled by the central nervous system.

Muscles working under any restriction, such as tight elasticated stocking or sock tops or garters, do not get adequate blood supply; waste products from activity, i.e. lactic acid, accumulate in the fibres and cramp results. This does not apply to 'support' stockings. A muscle if held too long in full contraction is depleted of its blood supply and will cramp (muscle spasm and pain); stretching or rubbing of the muscle or limb will assist the return of the blood flow.

Contraction of a muscle or muscle group is always followed by a phase of relaxation. Movement may be defined as alternating contraction and relaxation of muscle and the relaxation phase is of great importance to the efficiency of the muscle.

The work of a muscle is affected by the following facts:

a) A warm muscle works better than a cold muscle, e.g. warming up in track suits or woollies before intense activity.

b) If the muscle is first stretched, the following contraction is stronger and more efficient, e.g. the preparatory backswing in golf or tennis strokes.

c) The demand made upon it, related to 'All or None' law.

d) A fatigued muscle works less efficiently and is more prone to strain.

e) The blood supply must be free and adequate.

In order to work at all, a muscle must have a motor nerve to activate it, and a sensory nerve to inform the brain what is happening to the muscle or limb. The heat produced by working muscles produces a warming up of the body as a whole; this should be remembered when planning a class.

Muscle group action

Individual muscles combine to form muscle groups. Movements are performed by muscle groups; individual muscles are rarely, if ever, used, e.g. the muscle bulk on the front of the thigh works as a group to straighten a bent knee; the group, the quadriceps, is composed of four individual muscles having the same action (*see* Fig. 1/5). If one of the muscles in the group becomes inefficient, i.e. torn or bruised, the other muscles in the group will straighten the knee but the movement will be less strong.

In the body, muscle groups are paired and complementary; they are arranged on opposite surfaces of a

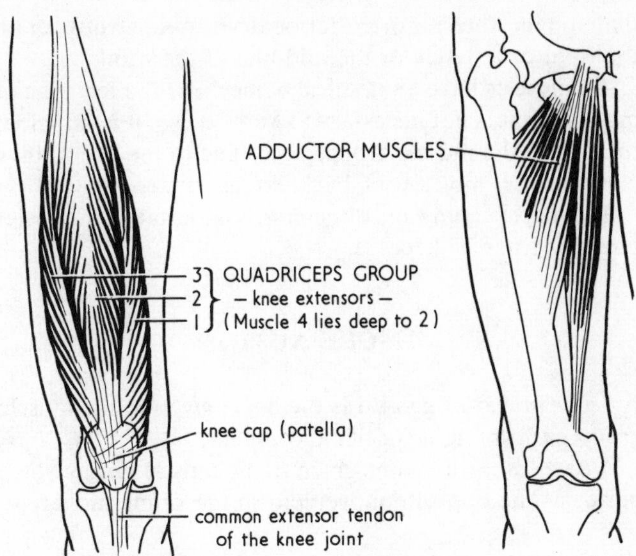

Fig. 1/5 Diagram to show the arrangement of the muscle groups of the thigh

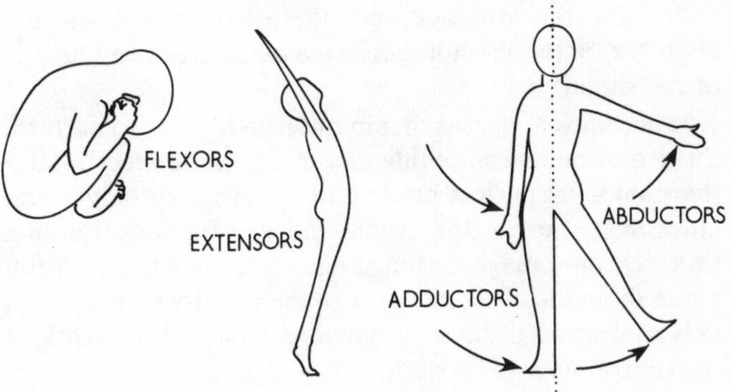

Fig. 1/6 Diagram to show the main muscle actions

limb, on the anterior or posterior aspect of the trunk, or on the right or left side of the mid-line of the trunk.

All muscles have anatomical names. For the teaching of movement it is not necessary to know these; it is sufficient to refer to the muscle groups according to the action they perform, e.g. hip flexors, back extensors (*see* Fig. 1/6).

The more reading and discussion is undertaken the easier terms and names become.

MUSCLE ACTION

A *prime mover or agonist* is the name given to the muscle group performing a specific movement.

Antagonist is the name given to the muscle group which performs the opposite movement to the prime mover.

Examples:

To flex the hip joint in order to go upstairs, the *prime movers* are the hip flexor groups; the *antagonists* are the hip extensors.

To raise the arm sideways, the *prime movers* are the abductors of the shoulder; the *antagonists* are the adductors of the shoulder.

It becomes apparent that in order to be able to perform these movements smoothly and in a co-ordinated fashion there must be perfect balance between prime mover and antagonist, i.e. as the prime mover contracts (getting shorter), the antagonist must allow a decrease in tension equal to and at the same rate as the activity of the prime mover. This co-ordinating function is part of the work of the central nervous system.

Muscle groups may also fulfil another function, that of fixing or holding part of the body while another more distal part is working, e.g. the muscle group around the shoulder holds the arm steady above the head while the hand and fingers replace a light bulb, draw curtains and so on. This method of working as a group is known as '*fixator*' work. Sometimes the work of a fixator is very strong and too tiring and often unsuitable (*see* Fig. 1/7).

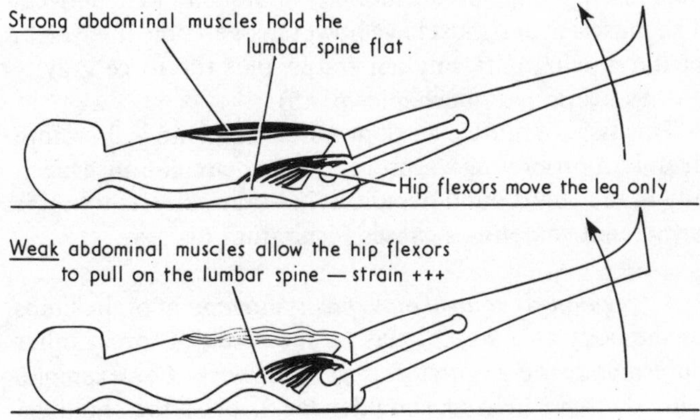
Strong abdominal muscles hold the
lumbar spine flat.
Hip flexors move the leg only
Weak abdominal muscles allow the hip flexors
to pull on the lumbar spine — strain +++

Fig. 1/7 Diagram to show the action of 'fixator' muscles

When planning movement sessions all forms of muscle work must be considered, not only that of the moving part, which is frequently the easiest, e.g. hand and wrist movements with the arm held in one position.

The function of muscles in action and some of the effects produced

1 To produce movement
2 To regulate or control movement
3 To hold positions produced by initial movement

1 *To produce movement*: the working muscles increase in size or bulk, become harder and shorter, drawing the two ends of the muscle nearer to one another. In practice usually one end is fixed (held) and the other end moves towards it, e.g. when flexing the elbow, the forearm usually moves towards the upper arm, not upper arm to hand. The ability to produce movement depends on the available power of the muscle group and the mobility of the joint to be moved. The muscle group must have power to overcome the weight of the moving part, any apparatus, plus the force gravity always exerts (*see* mechanics p. 45).

This type of muscle work produces heat and is therefore useful for promoting warmth and better circulation, especially in the chairbound person. It has other effects too; it is tiring, and sometimes causes respiratory distress.

2 *To regulate or control movement*: movement of the limbs or the body as a whole, may be the result of forces other than muscles, e.g. gravity or partner work. For example, with the arm held at shoulder level, once the 'holding' muscles relax, the weight of the arm will yield to the force of gravity and the arm will drop to the side; the speed at which it falls can be controlled by muscle power. To sit slowly on to a chair, once the initial movement is made to bend at the hip joints, the weight of the person plus the force of gravity is responsible for the action, but the speed and control of sitting is voluntary muscle work. In both cases, the controlling muscles are contracting, but at the same time they are being lengthened by other forces, i.e. gravity + weight.

In this type of muscle work the muscle is in a shortened or contracted state at the beginning of the movement, and during the movement the two ends of the muscle are drawn away from one another.

Less effort is required for this type of work; it produces

less heat but it too can cause breathing problems if the required control is strong. It is useful in the training of co-ordination.

3 *Muscle groups used to hold positions*: this is also known as 'static' muscle work. Having achieved a position by voluntary muscle action, the same muscles can hold that position by maintaining their state of contraction, i.e. no change of muscle length occurs. This form of work is tiring, tends to make people hold the breath and hampers the blood flow through the working muscles. The longer the 'hold' the more deficient the blood flow becomes and pain and cramp may develop.

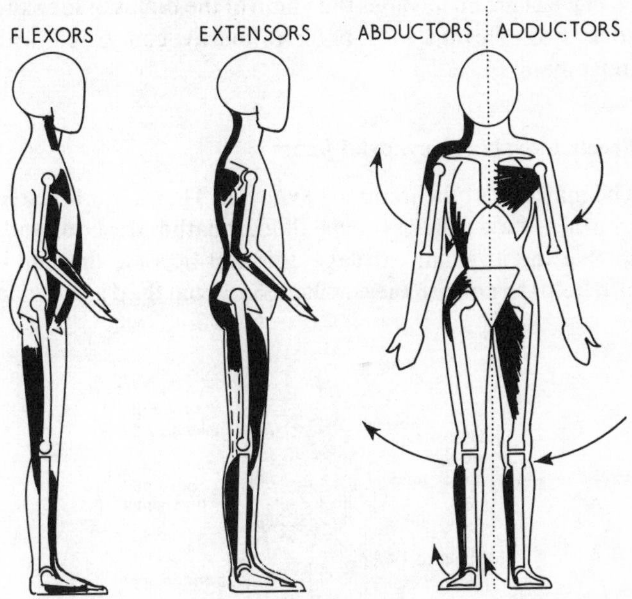

Fig. 1/8 Diagram to show the main muscle action groups

A limited amount of static work is useful in training posture and functional movement, but it should be brief and interspersed with good activity.

ARTICULATIONS or JOINTS

An articulation or joint is formed where two or more bones meet. The articulation formed may be:

1 A freely movable joint.
2 A semi-movable joint.
3 An immovable joint.

The last group includes the union of the bones of the skull so is not relevant in a book primarily concerned with movement.

Freely movable or synovial joints

The majority of the joints are synovial. They are lubricated by a clear fluid called synovial fluid. It bathes the bone ends so that the articular cartilage does not become dry and it also helps to nourish the cartilage. Synovial fluid is secreted

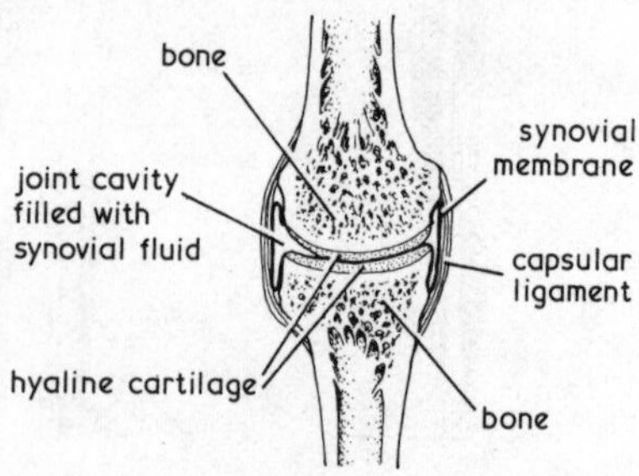

Fig. 1/9 Diagram of a synovial joint

by the lining of the ligamentous capsule of the joint – the synovial membrane.

The characteristic features of all synovial joints are shown in figure 1/9. Each individual joint has some modification to meet a special need at that particular joint, i.e. for extra stability.

FEATURES TO NOTE

1 The articular cartilage covers the adjacent surfaces of both bones.

2 A ligamentous capsule connects the bone ends, and is attached just beyond the articular cartilage. This is not a tight binding together.

3 Extra ligaments reinforce the capsule.

4 A synovial membrane lines the capsule and secretes synovial fluid to lubricate the joint.

The shape, size and direction of the articular surfaces determine the degree and type of movement possible as well as the plane in which the movement takes place. The larger the articular surfaces, the greater the range of available movement; this does not affect the actual mobility of the joint. Variations in joint mobility are obvious in individuals; some are naturally stiff, some hyper-mobile. Joint range decreases with advancing years.

Ligaments have a restraining effect on excessive movement at joints, particularly the extra ligaments already mentioned, e.g. extra ligaments in front of the hip joint to counteract the stress on the front of this joint by people who stand in a 'hip-sling' posture. The main support of any joint comes from the muscle groups which surround it.

Examples of freely movable joints:

The hip joint – a ball and socket type of joint.

The shoulder joint – also a ball and socket joint.

The knee joint – a hinge joint.

The elbow joint – a hinge joint; the rotating movement of the forearm occurring at the elbow region is a pivot joint.

A chart of joints and movements will be found in Chapter 2.

Semi-movable joints

These are joints at which a limited degree of movement only is possible. Semi-movable joints are found in the mid-line of the body. To have a fuller range of movement at these joints could endanger other vital structures near by.

Examples of semi-movable joints:

The joints between adjacent vertebral bodies.

The joint uniting the two pelvic bones in front.

The joint between the upper and lower part of the breastbone (sternum).

Vertebral joints are of two types, the semi-movable type between the bodies, and freely movable joints between small articular areas on lateral or transverse processes, projecting from the arch of the vertebra: these have all the characteristic features of synovial joints (Fig. 1/10).

The semi-movable joint between adjacent vertebral bodies is formed by the two bones and an intervertebral disc. The disc is made of fibro-cartilage. It varies in thickness, and responds to pressure whether even or uneven; pressure distorts its shape and creates movement at the joint (Fig. 1/11).

A disc is therefore compressible; it returns to its normal shape once the compressing force is removed; the compression on one side of the disc and the inevitable stretch of the opposite side of the disc, is very limited in degree, and is protected by strong ligaments.

Abuse or misuse of these joints results in the common back ache or even disc problems if the strain is severe.

The other semi-movable joints are not of real importance here.

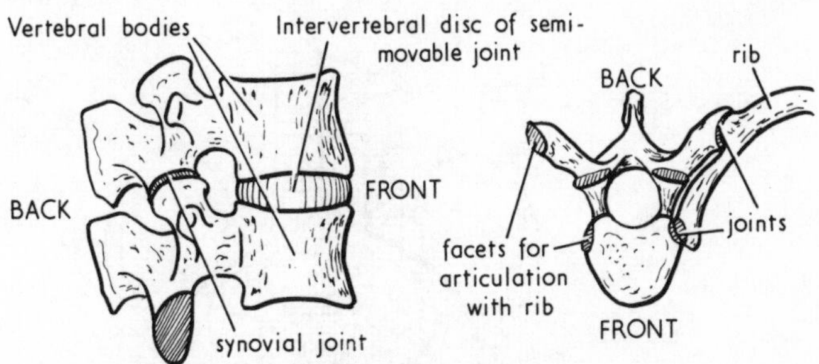

Fig. 1/10 Diagram to show semi-movable and synovial joints between articular facets of the adjacent vertebrae and between the rib and vertebra

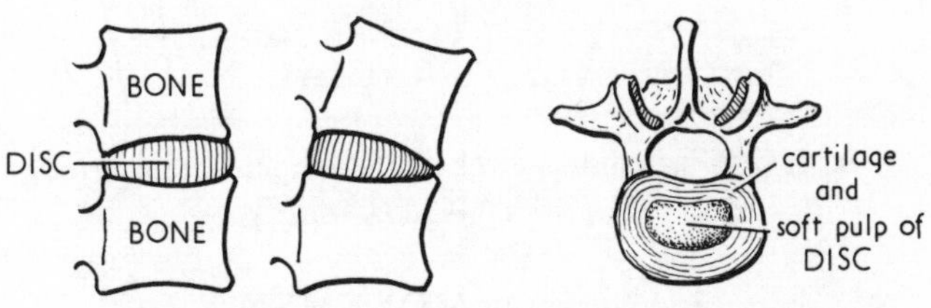

Fig. 1/11 Diagram to show the intervertebral disc

Immovable joints

True immovable joints are irrelevant. Two joints although classified anatomically as synovial, should be considered as immovable when considering exercise, Keep Fit or any form of free movement; these are the two sacro-iliac joints. The lowest part of the spine, i.e. the sacrum, articulates with the two pelvic bones at these joints, and should complete a secure bony ring, the pelvis. The body weight is transmitted to the legs via the pelvis which must obviously be strong and secure (Fig. 1/12).

Forced hip movements can strain this quite shallow joint.

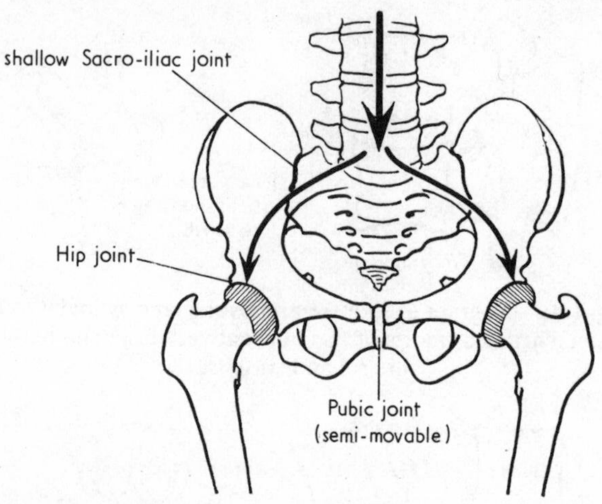

Fig. 1/12 Diagram to show how the weight is transferred through the pelvis

RANGE OF MOVEMENT

Joints and muscles move through a range of motion. For a joint, a full range is from one extreme of position to the other extreme, i.e. from full extension to full flexion. In order to maintain a fully mobile joint, movements which will use a joint throughout the whole range should sometimes be included in movement sessions.

Muscle range is measured from the fully stretched position of the muscle to its fully contracted or shortened position. In everyday life muscles or joints rarely work in the whole of their range, usually it is the middle of the range that is used, i.e. neither fully stretched nor fully contracted (Fig. 1/13).

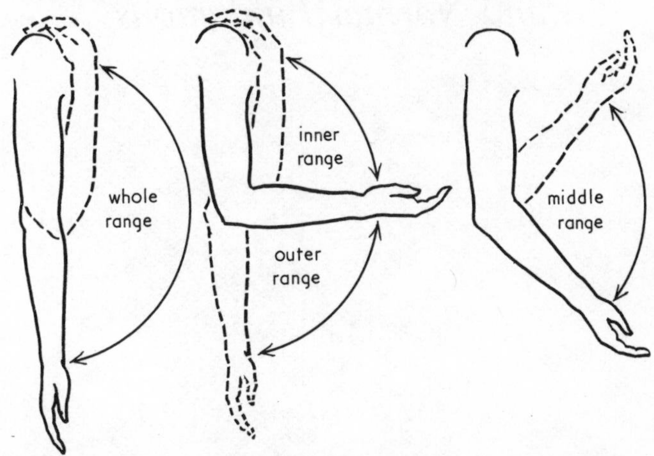

Fig. 1/13 Diagrams to illustrate the range of movement
Whole range = elbow bending, the hand to touch the shoulder.
Outer range = elbow bending to a right angle.
Inner range = elbow bending from a right angle to touch the shoulder with the hand.
Middle range = part of the outer range and part of the inner range.

Because of the anatomical arrangement of muscle groups on either aspect of a limb, the position of full shortening of one muscle group is the position of full stretch of the opposite group; both muscle groups therefore are adequately activated. Figure 1/13 illustrates the different ranges.

CHAPTER 2

Classification of Joints, Movement and Normal Limitations

JOINT	CLASSIFICATION	MOVEMENT	LIMITATION OF MOVEMENT OR SPECIAL NOTES
LOWER LIMB			
The hip joint	Synovial, ball and socket: ball = head of femur, socket = cup on lateral part of the pelvis.	Flexion	Limited by the opposite muscle groups especially the hamstrings if the leg is straight, or contact with the body if the knee is bent.
		Extension	Limited by the ligaments in front of the hip joint; after approximately 15° beyond standing position, the back arches, and the movement takes place in the lower spine at the same time the pelvis and raised leg move on the supporting leg.
	Fig. 2/1	Abduction	Limited by the tension of opposite muscle group and bony contact of the tip of the femur with the edge of the socket.
		Adduction	Limited by the opposite muscle group and contact with the other leg.

JOINT	CLASSIFICATION	MOVEMENT	LIMITATION OF MOVEMENT OR SPECIAL NOTES
The hip joint continued		Rotation outwards or inwards	Limited by the opposite muscle groups and strong ligaments.
		Circling movement	A combination of all movements with corresponding limitations.
The knee joint	Synovial, hinge joint: condyles of the femur, and the upper end of the tibia. Note – the outer bone of the lower leg, the fibula, is not included in the knee joint.	Flexion or bending	Limited by short quadriceps muscle or by the meeting of the calf muscle with the posterior aspect of the thigh.
		Extension or straightening	Limited by the limit of the articular surfaces and by tight hamstring muscles, if the hip is flexed. A small degree of extension, not normally used in standing, is available by voluntary contraction, pulling up the knee cap (patella).
		Lateral movement NIL	Prevented by very strong ligaments on the inner and outer sides of the joint.
	Fig. 2/2	Rotation outwards	No voluntary rotation occurs at the knee joint; a small degree of rotation outward of the tibia on the femur occurs mechanically at the final part of extension.

		Rotation inwards	The same applies to inward rotation of the tibia on the femur; it occurs at the initial stage of flexion. This mechanical movement is due to the structure of the joint.
The ankle joint	Synovial, hinge joint: a mortise is made by the lower ends of the tibia and fibula; this fits the upper saddle-shaped surface of the talus, a bone of the foot.	Flexion or dorsiflexion (foot pulled up towards knee)	Limited by the short tendo-calcaneum (tendo Achilles) and by the shape of the joint surfaces.
	Fig. 2/3	Extension or plantar-flexion (pointing the foot down)	Limited by the tendons and ligaments in front of the ankle joint.
		Lateral movement NIL	No active lateral movement. Passive movement sideways if the ankle is fully extended.
		Rotation NIL	For apparent movement see joints of the foot.
Joints of the foot a) mid-foot	Synovial: gliding joints of several small bones make up this complicated joint.	Inversion, or turning in	Pure movement only if the ankle is dorsiflexed. Limited by ligaments and tendons on the outer side of the ankle and foot, and the smallness of the joint surfaces.
		Eversion or turning out	Similar to inversion, but the limitations are on the inner side of the foot and ankle.

JOINT	CLASSIFICATION	MOVEMENT	LIMITATION OF MOVEMENT OR SPECIAL NOTES
Joints of the foot continued			
b) Between all small bones adjacent to one another	Synovial: gliding movement allows resilience of the foot.		These small bones, which form gliding joints, help to form the arch of the foot. The arch is supported by ligaments and muscles on the sole of the foot.
Joints of the toes	a) Synovial, hinge joints:	Flexion or curling up the toes	Limited by the limit of articular areas and can also be limited by short muscle tendons.
		Extension or turning the toes upward	Limited by the small articulations, and tendons on the sole of the foot.
	b) modified joints at the base of the toes.	Abduction and adduction	Very little movement of separating and closing the toes from the mid-line of the foot.
		Foot circling	This is a combination of movement at the ankle joint and of the joints of the mid-foot. It is composed of flexion and extension at the ankle joint and inversion and eversion at the mid-foot. The limitations are those for the individual joints.

UPPER LIMB

Note: Full shoulder movement relies on a full range of movement at the shoulder joint and at the joints of the shoulder girdle.

JOINT	CLASSIFICATION	MOVEMENT	LIMITATION OF MOVEMENT OR SPECIAL NOTES
The shoulder joint	Synovial, ball and socket: ball = head of the humerus; socket = very shallow area on the outer border of the scapula.	Flexion	Limited by the ligaments of the shoulder joint, and opposite muscle group.
		Extension	Limited by the ligaments and opposite muscle group.
		Abduction	Limited by the head of the humerus coming into contact with the bony arch of the shoulder girdle. This limit varies in individuals; once contact is made, the arm can only be raised above the head by moving the shoulder girdle.
	Scapula Humerus Fig. 2/4	Adduction	Limited by contact of the arm with the body and by the opposite muscle group and ligaments.
		Rotation outwards and inwards.	Limited by ligaments and muscle groups surrounding the joint. Because of the extreme shallowness of this joint, the muscles play a vital part in the security of this region.

JOINT	CLASSIFICATION	MOVEMENT	LIMITATION OF MOVEMENT OR SPECIAL NOTES
The shoulder joint continued		Arm circling	This is a combined movement of both shoulder joint and all joints of the shoulder girdle. The limitations are as for those of both joints.
The shoulder girdle This has three joints:		As a result of the gliding movement the scapula moves on the chest wall	The movement of the shoulder girdle can be independent of the shoulder joint.
a) sterno-clavicular	Synovial – the clavicle glides with the sternum in front.	These movements are: (*see* Fig. 2/5) A. Vertical, hunch and lower	Scapula muscles produce the movement.
b) clavicular – scapular	Synovial – the clavicle glides laterally with a process of the scapula (shoulder blade).	B. Abduction and adduction; the scapula moves round the chest wall or draws the scapula towards the spine	Limitation of movement will be according to the size of the articular surfaces and the available freedom of the scapula.

c) scapula (shoulder blade) on the chest wall		C. Rotation of the lower angle laterally and upwards *or* downwards and inwards	

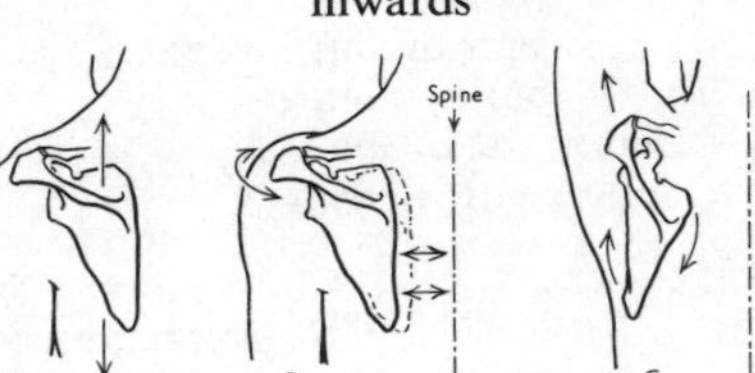

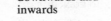

Fig. 2/5 Diagram to illustrate the direction of movement of the scapula on the chest wall.
A. Vertical, hunch and lower B. Abduction and adduction C. Rotation of the lower angle

The elbow joint	Synovial, hinge joint: formed by the lower end of the humerus and the upper end of the ulna which is the inner bone of the forearm.	Flexion	Limited by the limit of articular areas; ligaments and opposite muscle, and the contact of upper and lower arm muscles.
		Extension	Limited by the locking of the joint by the bony process of the ulna into the fossa of the lower end of the humerus (fossa = a deep indentation).
		Lateral movement NIL	Limited by the shape of the bony surfaces and strong ligaments.

JOINT	CLASSIFICATION	MOVEMENT	LIMITATION OF MOVEMENT OR SPECIAL NOTES
Forearm joints	Synovial, pivot joint: formed at elbow level between the head of the radius, which is the outer forearm bone, and the ulna.	Supination or turning the palm of the hand upwards. Pronation is the reverse	The radius pivots in a circle of ligament, attached to the ulna; the lower end of the radius actually crosses the ulna taking the hand with it.
The wrist joint	Synovial, condyloid type of joint – like an egg in a spoon; it moves in 4 directions only; being oval no rotation is possible.	Flexion, hand towards the forearm	Limited by the limit of joint surfaces and ligaments and tendons at the back of the wrist.
		Extension	Limited by the limit of joint surfaces, and structures in front of the wrist.
		Radial flexion, or movement of the hand to the thumb side	Limitation to both of these movements by size of joint surface, bony contact, and opposite ligaments.
		Ulnar flexion or movement of the hand to the little finger side	

		Wrist circling	This cannot occur as the joint is not a ball and socket type. A combination of wrist and forearm movements will result in a circling type of movement.
The joints of the hand taken as a unit	Synovial; a) mostly gliding joints at the 'heel' of the hand and between the bones which form the palm; b) the finger joints are similar to the toe joints and move in the same direction.	Flexion and extension; or (clenching and stretching)	Fingers have shallow joint surfaces and can be easily hurt or damaged if abused, e.g. with hard or heavy apparatus. Muscle tendons and ligaments protect flexion and extension movements. Ligaments are the only protection against lateral strain.
		Abduction and adduction to mid-line of the hand; or (parting and closing)	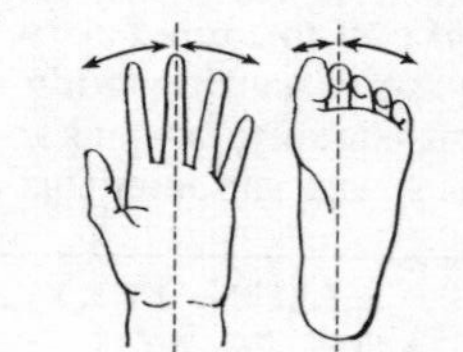 Fig. 2/6 Diagram to show the movement of the hand and foot in relation to its mid-line

The thumb has the special movement of opposition, i.e. the thumb can be placed across the palm of the hand, enabling the person to hold or grip, using the thumb round the object.

SPINAL JOINTS

The joints between adjacent vertebrae are of two types. The joint between two vertebral bodies is semi-movable and weight bearing; the joints between the small articular facets on the vertebral arches are synovial gliding joints. These latter joints determine the direction and degree of movement at a specific level of the spine.

JOINT	CLASSIFICATION	MOVEMENT	LIMITATION OF MOVEMENT OR SPECIAL NOTES
Joints between the vertebral bodies	Semi-movable.		A fibro-cartilaginous disc is adherent to the lower surface of one vertebral body and the upper surface of the body below. Strong anterior and posterior ligaments run the length of the vertebral column. The posterior ligament is within the vertebral canal.
		Flexion Extension Side flexion	Movement between vertebral bodies is due to compression of the disc, anteriorly to flex the spine, posteriorly to extend, laterally to side flex. The extent of movement is limited by the thickness of the disc and the opposing ligaments. Deep spinal muscles are the most important defence. The spine will return to the straight position once the distorting force is removed.

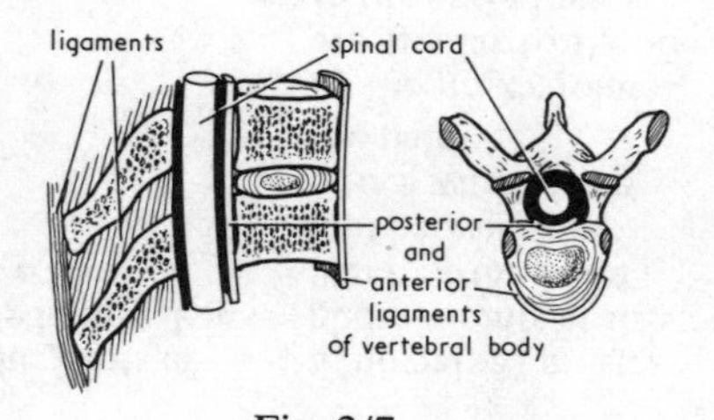

Fig. 2/7

Joints between the articular facets on the arches of the vertebrae	Synovial, gliding joints.	See notes below	These joints have all the features of typical synovial joints, and their limitations. Different planes of movement are possible in different areas of the spine, due to the direction of the facets. The range of movement at one joint is very small, but in total the range of spinal movement is considerable.
Cervical spine joints			
a) atlanto-occipital		'Nodding'	This is a special joint between the first cervical vertebra and base of the skull.
b) atlanto-axial		Pivoting or head turning	This is the joint between the first and second cervical vertebrae.
c) cervical joints 2–7		Flexion, extension, some side flexion, some rotation	Movement of the cervical spine is limited by the size and direction of the facets, the vertebral arches and by muscles and ligaments.
Thoracic spine joints		Flexion	Limited by the rib cage, the shape of the thoracic curve and the thin discs.
		Extension	Limited by the rib cage and the very oblique vertebral spines in this region.
		Side flexion	Limited by the structure of the rib cage.

JOINT	CLASSIFICATION	MOVEMENT	LIMITATION OF MOVEMENT OR SPECIAL NOTES
Thoracic spine joints continued		Rotation	This is possible throughout this region but is most free in the lower part – T7–T12.
Lumbar spine joints		Flexion	Limited by ligaments, strong back extensor muscles and the concavity of the lumbar spine.
		Extension	This movement is fairly free; limited by small articulations and ? the contact of vertebral spines.
		Side flexion	Limited by ligaments, muscles and size of articular facets.
		Rotation NIL	There is apparent movement which occurs in the thoracic spine above and the pelvis moving on the two femora, but the direction of the facets prevents rotation.

CHAPTER 3

Mechanics of Movement

Movement is produced by two factors:

a) by muscle power, already discussed.

b) by external forces, i.e. gravity or mechanical forces such as apparatus or a partner.

GRAVITY

Gravity is the force attracting all objects towards the earth. As this is a constant force it has always to be considered, whether there is motion or stillness.

Movement can be in the same direction as gravity, e.g. relaxing or bending the body forward, from the standing or sitting position.
Movement can be against the force of gravity, e.g. raising the body to the upright position again.
Movement can be performed with the effective force of gravity neutralised or balanced by some support, e.g. leg parting and closing while sitting on the floor with the legs fully supported.

Gravity is only one of many factors to be considered when planning any movement; these will be summarised later when individual factors have been considered.

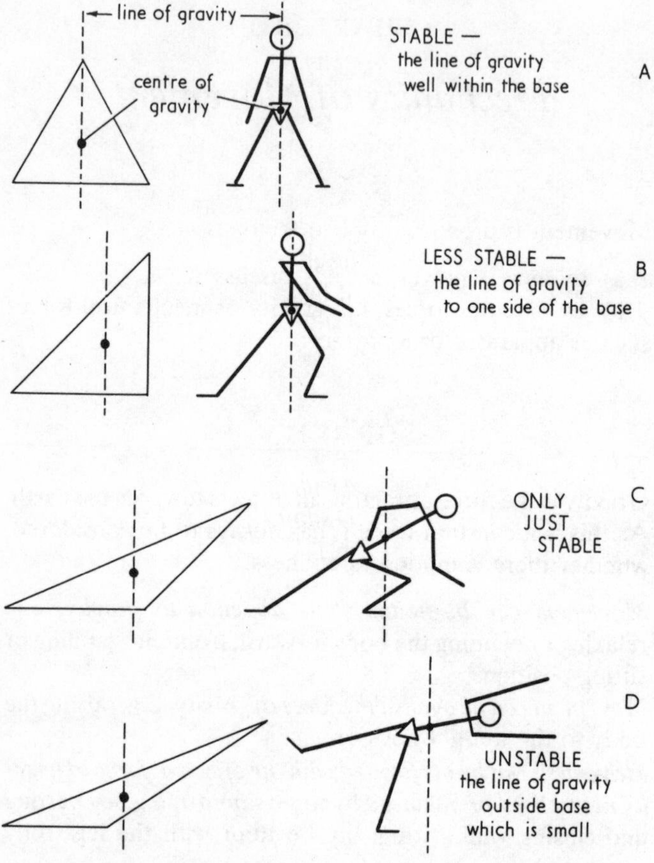

Fig. 3/1 Diagram illustrating the relationship between gravity, base and stability
---- = line of gravity ● = centre of gravity

Centre of gravity

The centre of gravity of a person or object, such as a limb, is 'that point through which the force of gravity passes to earth, and balance can be maintained'. In the human body the centre of gravity is said to be approximately at the level of the second sacral vertebra (Fig. 3/1). The centre of gravity varies with the position of the body, high when standing, lower when kneeling or lying. This variation affects balance and stability. Other factors affecting stability are the size of the base and the line of gravity in relation to the base.

Line of gravity

This is an imaginary line passing vertically through the centre of gravity to earth, it is important because of its relationship to balance (*see* Fig. 3/1).

Base or support

The base is the area of support. It includes not only the points of contact with the floor, but also the area within these points, i.e. standing with feet together, or with feet apart (Fig. 3/2).

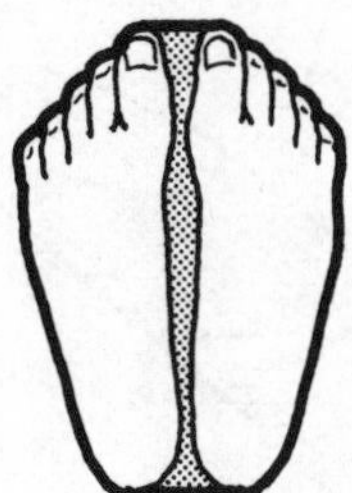

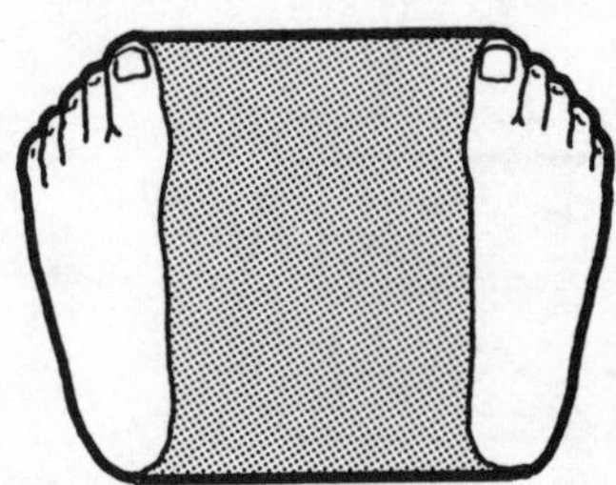

Fig. 3/2 Diagram illustrating area of base

Although the contact is the same, the area covered is greater with feet apart, and is therefore the more stable position. Sitting on a chair with feet on the floor, the base is large and the position stable; if the feet do not reach the floor the thighs represent the base and the position is far less stable.

The line of gravity in relation to the base is important. The nearer the line of gravity falls to the centre of the base, the more stable is the person or object (Fig. 3/1A). As the line of gravity falls nearer the perimeter of the base, stability decreases (Fig. 3/1B&C). If the line of gravity falls outside the base, balance or stability becomes precarious or impossible (Fig. 3/1D).

A wide base is advisable for people who are less mobile or co-ordinated, but imposes certain limitations in the movement which may become more formal and stereotyped – but safe. Another consideration is the shape and

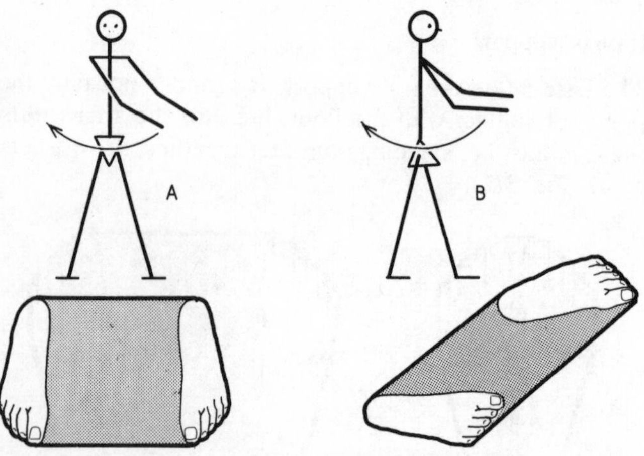

Fig. 3/3 Diagram to illustrate the direction of base in relation to movement

direction of the base in relation to the movement. Figures 3/3A&B have the same area of base but different direction. Figure 3/3A is more suitable for side-to-side movement; Figure 3/3B is better for forward and backward movement. Body weight is kept as much as possible within the base. Arm swinging forwards and backwards from position Figure 3/3A tends to overbalance the body, unless there is a compensatory swing to and fro from the ankles, which spoils the movement. Position Figure 3/3B is ideal for forward and backward swings, but of little help in sideway swings. Shape and freedom of movement are closely associated with base, line of gravity and balance.

INERTIA

The body or parts of the body are subject to inertia. This is the resistance of a body to a change, i.e. if still, it tends to remain still until some force moves it. Once moving it opposes a change either to stop or to change direction. Example, the weight and inertia of an arm has to be overcome by muscular effort in order to start arm swinging sideways, but once started, the movement can continue with very little additional effort. To stop suddenly or change the direction of the swing indicates the resistance to the change.

This fact is applicable to the elderly or stiff 'movers' who have difficulty in starting movement, and who may be thrown off balance by sudden stops or changes of direction.

LEVERS

The principles of levers and leverage can be used to modify movements by making them easier or more difficult.

Definition. A lever is a straight rod capable of rotation about a fixed point if acted on by opposing forces.

Levers are classified according to the relative position of three factors along the lever (*see* Fig. 3/4).

The three factors are:

1. the fulcrum, or fixed point about which the movement takes place.
2. the weight to be moved.
3. the effort, or point of application of the power to move the given weight.

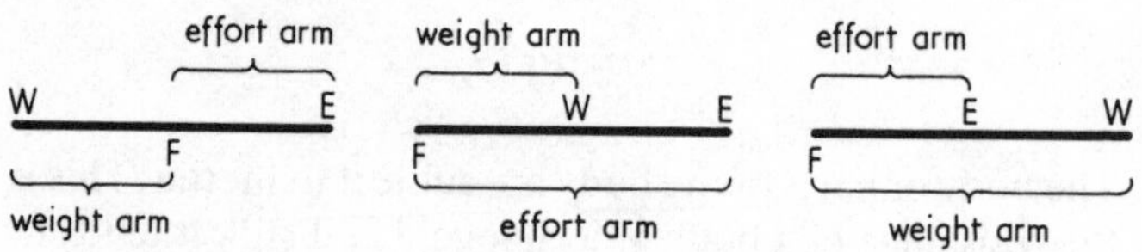

Fig. 3/4 Diagram to illustrate the three types of levers

The distance between the fulcrum and the weight is called the weight-arm; the distance between fulcrum and application of effort is the effort-arm.

The examples shown in Figure 3/5 assume that the lever is of uniform weight throughout. As in Figure 3/5A, with F placed centrally between W and E the bar will balance (say 4ft to either side of F).

In Figure 3/5B_1 as with a see-saw, if extra weight is added to either side, the weighted side goes down, the light side up. (Weight or effort are interchangeable.) The same will occur if one of the arms is increased in length (Fig. 3/5B_2).

In Figure 3/5C to balance, with uneven length of weight-arm and effort-arm, the required amount of effort may be calculated as follows:

Weight × length of weight-arm = effort× length of effort-arm.

e.g. 4lb × 2ft = unknown effort × 6ft. Therefore the unknown effort = $\frac{8}{6}$ = $1\frac{1}{3}$lb. In other words 8lb can be balanced by an effort of $1\frac{1}{3}$lb and this is an economy in terms of effort. To lift the weight possibly 2lb effort would be very adequate.

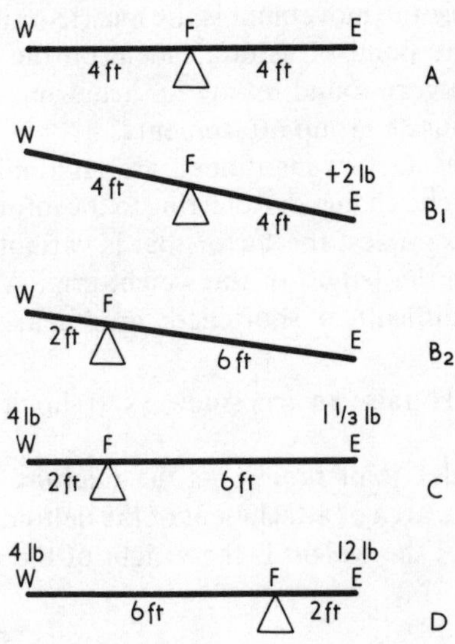

Fig. 3/5 Diagram to illustrate the principles of leverage

Figure 3/5C shows that when the effort-arm is always longer than the weight-arm, i.e. a mechanical advantage, less effort is needed in relation to the weight to be moved. Figure 3/5D shows the reverse; the weight-arm is the longer and requires more effort in relation to the weight to be moved, i.e. a mechanical disadvantage. Unfortunately most levers in the body are of this type, but there are some advantages such as speed or a greater range.

Application of 'Levers' to movement

In the human body the joint at which the movement takes place acts as the fulcrum. The weight is the part of the body or limb to be moved. The effort or power to move the part and produce the movement is the muscle or muscle group, taken at the point of its attachment on the bone (lever). This is one very sound reason for acquiring a basic knowledge of muscle group attachments.

The three factors mentioned are anatomical facts and they cannot be changed. Referring to the information in the previous examples, the factor that is variable and can be modified is the length of the weight-arm. A longer 'arm' makes it difficult, a shortened 'arm' makes movement easier.

Example: To raise an arm sideways to shoulder level (Fig. 3/6).

The shoulder joint represents the fulcrum, the effort or power is the area of attachment of the deltoid muscle, both unalterable; the weight is the weight of the arm.

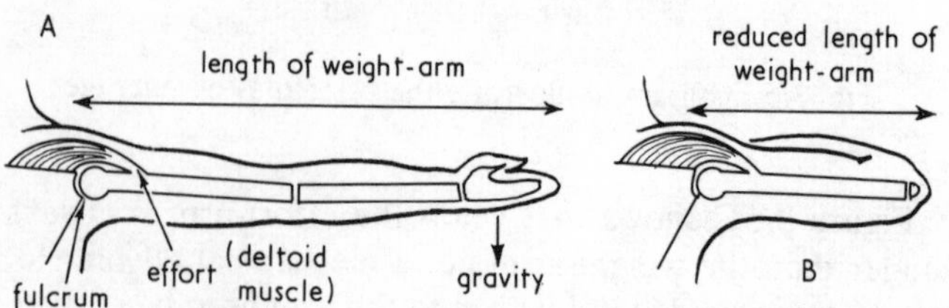

Fig. 3/6 Diagram illustrating how the arm is raised to shoulder level and showing how the weight-arm is modified

The arm weighs the same whether straight or bent, but by bending the elbow the weight-arm is shortened by almost half, therefore less effort is needed to raise it. Similarly, raising a straight leg is much harder work than raising a bent leg; in this example additional muscle work is involved.

Many movements can be modified by applying the principles of levers, by decreasing the weight-arm, as above, or increasing the work by increasing the actual weight, or increasing the length of the weight-arm by the addition of apparatus.

Friction

Friction is the resistance offered when two surfaces move on one another. This can be useful, essential or a hindrance, e.g. a slippery shoe on a slippery floor gets no purchase when walking and a person tends to fall, similar to walking on ice. Someone who has to shuffle along with rubber soles on a thick carpet has too much friction and can hardly move. A good rough sole on a non-slip surface gets a firm purchase and walking is easier. Resin is often used to provide friction for active movers on polished floors. Loose mats or rugs are dangerous, not only because they slip but also because they may cause the less stable to trip. Plastic soles are not always satisfactory; stocking feet should never be permitted; bare feet are safe and useful at times, but not always practical or possible.

TONE, TENSION AND RELAXATION

Tone

Normal resting muscle is neither flaccid (flabby), nor hard and tense; it has a quality of firmness, a texture that is pliable and resilient and gives shape. This state of muscle is known as the 'tone' of the muscle, it has nothing to do with bulk or weight; emaciated muscles have no tone.

Muscle tone is not like muscle contraction for it is not possible to alter tone voluntarily. Tone is controlled by the central nervous system; tone is defined as the state of readiness of muscle for action or work.

Resting muscle is not in a state of permanent contraction, although it is in a state of good tone; there is no slackness in the length of the muscle which has to be taken up before movement can commence. A child's toy on a string will not pull along until some effort has been used to tighten the string first (*see* Fig. 3/7).

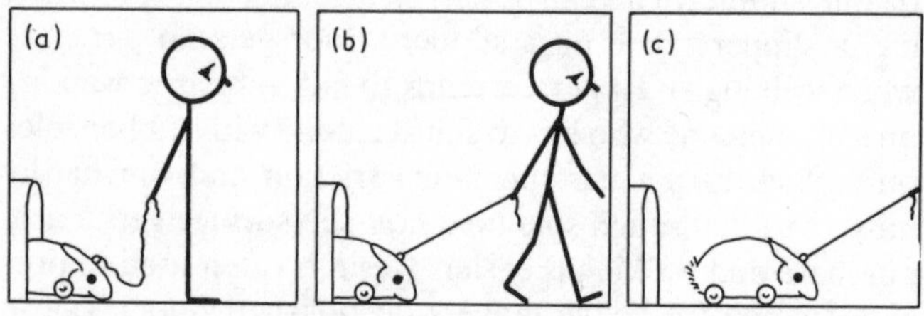

Fig. 3/7 Diagram to illustrate muscle tone

Muscle corresponds to b) and produces activity on the first stimulus, c).

As age advances muscle tone diminishes; the once firm muscle becomes less firm and the contour or shape of limbs and muscle groups changes as the tone is lost. The effects of this diminution of tone are more than just cosmetic. Joints once protected by firm tonic muscles become more prone to strain, and initiation of movement is slower. Loss of tone does not necessarily mean loss of power, but is frequently associated with it.

Tone must not be confused with tension. Tension may be the existing state of muscle in some people, but is not normal. The balanced tone of muscle groups plays a major part in the maintenance of posture. Upright posture could not be held by constant active contraction of various postural muscles interacting with one another, and against outside forces such as gravity. Such a situation would be too exhausting, and would require constant conscious effort. In some instances muscle tone can be improved without increasing bulk or creating tension.

A period of inactivity, e.g. illness, results in loss of tone; a gradual return to normal activity helps to regain tone and postural efficiency.

Tension

Tension may be described as a state of excessive muscle tone. It is not a normal state and creates many problems. Certain types of individuals are tense in themselves, and their appearance is usually sufficient to single them out. Tension is also controlled by special areas of the central nervous system – one area for excitability, one to inhibit or dampen down; if these are balanced then an even tone will result. In states of tension, the inhibitory area is under-active, leaving excitable muscle predominant.

Tense muscle do not get adequate rest periods, the blood supply is restricted, and headaches, postural pains, cramp, etc. are the result.

Associated with physical tension is mental stress or anxiety of which the individual may or may not be aware. Attendance at some sort of group for physical activity is a useful outlet as a temporary measure. A relaxing type of movement is a great help, as physical and mental tension being closely associated, one affects the other. The ability to recognise a state of tension is part way to reducing it.

Relaxation

Relaxation may be considered as the opposite to tension. Some people relax easily; the relaxed person tends to take things as they come and can generally find some humour in most situations.

The majority of people are less fortunate and can only relax under ideal conditions. To these people and to the 'tense' group it is consoling to know that relaxation is an acquired skill and with perseverance and practise some success can be achieved if the efforts are sincere. As for all skills, self-discipline is necessary to give up time regularly to practise.

The ideal conditions mentioned are physical comfort, i.e. warmth, ease of mind, or a sound and satisfied digestion. Those who declare they can 'relax on a clothes-line' will have difficulty if faced with adverse conditions, as they probably have no positive skill to apply. Those who have to work to achieve this skill have an asset to use in a diversity of situations.

Relaxation is the releasing of tension and it should be studied or taught as a positive activity, a 'do' rather than a 'don't'. It is useless to tell a tense person to 'relax'; they do not know what it feels like, for to them it is not a conscious recognisable state. As with the degrees of muscle control, awareness must play its part. Therefore time must be allowed, to register the difference in quality between relaxation and tension.

Relaxation can be local or general and there are many ways of teaching. The frequent euphemism 'think of pleasant things, lying in the sunshine' does not really meet the need of the executive official with tension headache, or the housewife with domestic problems and loss of sleep; to them such abstract things are only an irritant. The approach must be positive, concrete and active. It must be understood that success does not come overnight.

Relaxation does not necessarily produce sleep, but it does produce rest. The systems of the body work at a minimum, i.e. no muscle tension, the circulation is slowed down and respiration is reduced to minimal rate and depth.

PRINCIPLES INVOLVED IN THE TEACHING OF RELAXATION

1. Support; to ensure that no muscular activity is needed to support the body or part of it.

2. Physical comfort, i.e. ventilation, warmth, subdued lighting and as little noise as possible at first, is advisable to aid concentration.

3. Individuals must be free to move if uncomfortable; just to be still is very trying to a tense person.

4. The eyes may be kept open as sleep is not anticipated; confidence in the teacher is essential.

5. Stretching limbs or spine, not flexing, and then releasing the tension often produces the best results. Swinging, pendular types of movement and slow breathing exercises in personal timing, can also be used as an aid to relaxation.

CHAPTER 4

The Respiratory System

The respiratory system is often forgotten until it makes itself felt or its efficiency breaks down. A very brief outline of the system and its function follows. No one system of the body can be entirely isolated for in some way each affects every other system, and deficiency of one will affect the efficiency of all the others, e.g. deficient oxygen will affect the nutrition of all tissues of the body.

THE ANATOMY OF THE RESPIRATORY SYSTEM

The respiratory apparatus consists of a passage way from the nose via the back of the throat (pharynx and larynx) to the trachea or windpipe (Fig. 4/1). This becomes the main bronchus in the upper part of the chest. Small hairs in the nasal passages help to filter dust from the air and prevent it reaching the lungs and moisture is obtained from the mucous lining of the airway as well as from the atmosphere. The trachea can be felt in the front of the neck and is kept open by cartilaginous rings. The main bronchus runs through the centre of the upper chest, then divides into a large right and left bronchus, one to each lung. Each bronchus divides into smaller and smaller branches; these terminate as minute bronchioles in the air sacs of the lungs. The air sacs are called alveoli. This branching system is known as the bronchial tree.

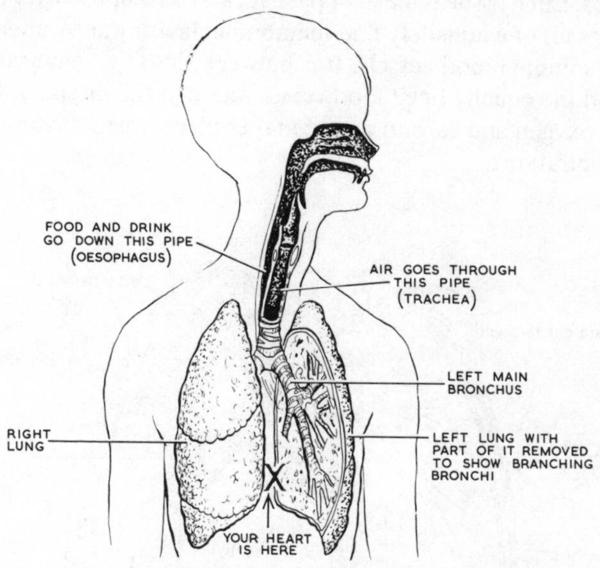

Fig. 4/1 Diagram illustrating the respiratory tract

The lungs are triangular in shape; the base rests on the diaphragm muscle and the apex can be felt at the base of the neck if a very deep breath is taken. Both lungs are adherent to the inner walls of the rib cage via the pleura (*see* p. 61). They fill the whole chest cavity except for a central area containing the heart, the main blood vessels and the oesophagus; the latter passes from the mouth through the thorax and the diaphragm to enter the stomach.

Each lung is divided into lobes; three lobes for the right lung, two lobes for the left lung. The pleura covers each lobe (Fig. 4/2). Breathing, i.e. inspiration and expiration, is the result of mechanical movement of the thoracic cage as well as through changes in volume and pressure in the air sacs. Lung tissue is inert, or passive, and is composed of tiny air sacs of exquisitely fine membrane, lavishly surrounded by minute blood vessels. It is between this fine membrane and the equally fine blood vessel wall that the interchange of oxygen and carbon dioxide takes place; this is 'external respiration'.

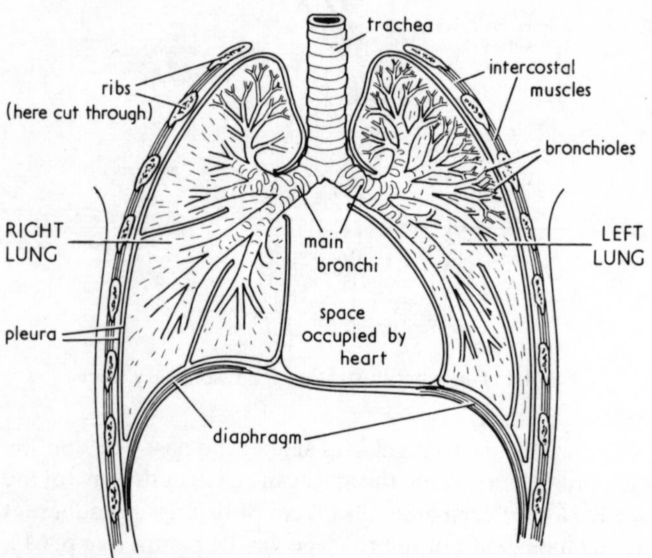

Fig. 4/2 Cross-section of thorax to show pleura and divisions of the bronchial tree

The thorax is somewhat restricted in its movement because of its contents.

The thorax is made up of the thoracic spine, twelve pairs

of ribs and the sternum (breastbone). The spaces between the ribs are filled by intercostal muscles and the floor of the cage is formed by the diaphragm muscle (Fig. 4/2). The joints between the ribs and vertebrae are synovial and freely movable, although the range is small. Anteriorly, cartilage joins the rib to the sternum (ribs 1–7), while the cartilages of ribs 8–10 join the cartilage of the rib above; this gives the shape of the costal margin. Ribs 11 and 12 are called 'floating ribs' – this is a bad description for they are very firmly joined to the vertebrae and provide attachment for many muscles, as well as helping to protect the kidneys.

The sternum is relatively fixed although the joints between sternum and the rib cartilages are semi-movable.

In order to enlarge the chest cavity when breathing, movement occurs in all directions. The thoracic spine extends slightly, the sternum moves forwards and upwards, thus increasing the antero-posterior space; the ribs move sideways and upwards, like a bucket handle, thus increasing the lateral dimension; the diaphragm descends, increasing the vertical space. Because they are less fixed, the lower ribs are more mobile than the upper ribs.

The muscles involved in inspiration are the diaphragm and the muscles between the ribs; in very deep breathing, muscles in the neck are also used to raise the upper ribs; the abdominal muscles should relax during inspiration to allow the diaphragm to descend with minimal resistance. Expiration is entirely passive during quiet breathing; in forced expiration the abdominal muscles are brought strongly into action to compress the abdominal organs, preventing the descent of the diaphragm and mechanically reforming its dome.

Pleura. Each lung is enveloped in a double fold of membrane, the pleura (Fig. 4/2). The inner layer is adherent to the lung, the outer layer to the inner walls of the thorax and the upper surface of the diaphragm. There is no true space between the two layers. It can be understood that any movement of the thorax must be accompanied by stretching or compression of the underlying lung.

PHYSIOLOGY OF RESPIRATION

Physiologically, respiration means 'interchange of gases'. It can only occur in that part of the bronchial tree which is capable of carrying out this function and in the human body this includes the air sacs (alveoli), the minute bronchioles and the blood vessels (Fig. 4/3). The upper part of the respiratory passages only conveys the air to the lungs.

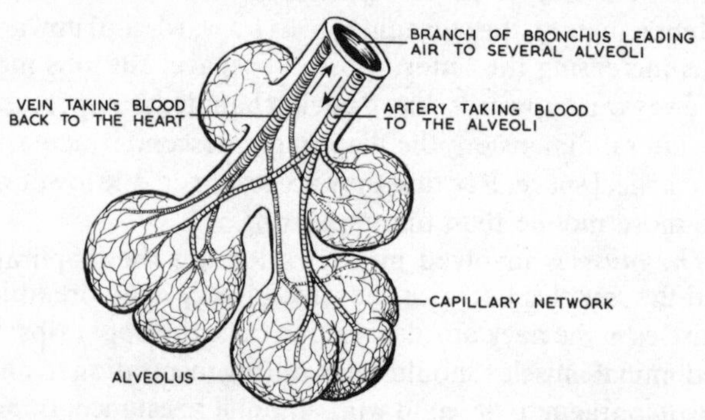

Fig. 4/3 Diagram of the alveoli (air sacs), showing the blood vessels which carry out the gas exchange

The gases are exchanged between the air in the alveoli (air sacs) and the tiny blood capillaries in contact with them – this is external respiration. Gases are also exchanged between the blood, which contains gases in solution, and the body tissues – this is internal respiration.

The Respiratory System

In general terms respiration is interpreted as 'breathing' and the normal rate is approximately 18–20 breaths per minute.

Inspiration = intake of air;
Expiration = expulsion of air;
True respiration is the unseen interchange of gases.

Internal respiration and the capillary bed

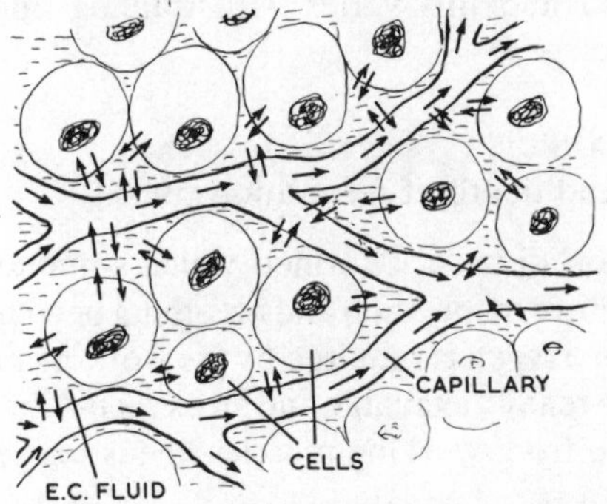

Fig. 4/4 Diagram showing the capillary bed and the process of internal respiration. E.C. = extra-cellular fluid

Oxygen in solution at the arteriole end of the capillary bed is held until capillaries are formed. The capillary walls are so thin that fluids are able to pass through them to the tissue cells via the tissue fluid. Because of a difference of pressures, fluid containing waste products is drawn back into the venous end of the capillary bed; once the veins begin to form, interchange of fluids ceases and internal respiration can no longer take place.

Although breathing can be controlled voluntarily for short periods, it is basically under the control of the respiratory centre in the brain.

CHANGES IN RESPIRATION RELATED TO EXERCISE

ANTICIPATORY

This is psychological. The respiration rate increases and therefore the rate of the circulation through the lungs is increased. There is an increase in the venous return to the heart; the skeletal muscles receive more blood and oxygen and are ready for action. The depth of respiration may be affected and this varies with training and the work anticipated.

DURING EXERCISE

The rate and depth of respiration rises:

a) carbon dioxide is formed which stimulates the respiratory centre in the brain and breathing becomes deeper.

b) more oxygen is required by the working muscles; this may not be readily available and an oxygen debt is built up.

c) waste from working muscles needs oxygen to eliminate it.

d) body temperature rises and this also leads to an increase in respiratory rate.

AFTER EXERCISE

The respiratory rate and depth return to normal, the oxygen debt is repaid, carbon dioxide is removed via the lungs, and waste products, e.g. lactic acid are eliminated via the blood stream. The length of recovery time depends on the fitness or training or the individual.

TERMS ASSOCIATED WITH RESPIRATION

Anoxia Inadequate supply of oxygen reaching the tissues. There are several causes for this condition.

Cyanosis This is the name given to the blue tinge seen on the lips, ears or extremities and it is due to poor oxygenation of the blood. It may be seen in some chronic chest conditions.

Dyspnoea The condition of difficult breathing; panting after little or no exertion. It may be due to heart conditions, asthma or as a temporary, normal condition, e.g. at high altitudes or after strenuous exercise, although the latter hardly comes under this heading.

Emphysema This is a condition of the airways particularly affecting the alveoli, which lose their elasticity becoming more stretched and allowing for less space through which exchange of gases can take place (*see* Fig. 8/4, p. 101). The person with emphysema will be breathless.

CHAPTER 5

The Circulatory System

The circulatory system can be described in two parts:

1. The systemic circulation (Fig. 5/1A).
2 The pulmonary circulation (Fig. 5/1B).

Both systems start and finish at the central pump, the heart. The heart is responsible for discharging blood through a closed circuit via a system of arteries and veins, to the various parts of the body. The heart receives the blood from the body in readiness for the next circuit.

THE ANATOMY OF THE CIRCULATION

The heart is a triangular shaped organ with the base uppermost and lying mostly to the left of the mid-line of the chest. It is divided into four chambers; the right and left atrium (upper chambers), and the right and left ventricle (lower chambers) (Fig. 5/1B). The two right-side chambers communicate, and likewise the two left-side chambers communicate, but there is no direct communication between the right and left sides of the heart. The walls of the heart are made of cardiac muscle. This is a specialised muscle characterised by its property of continuous rhythmical contraction and relaxation. The rate and strength of the contraction is controlled by a cardiac centre in the brain. This

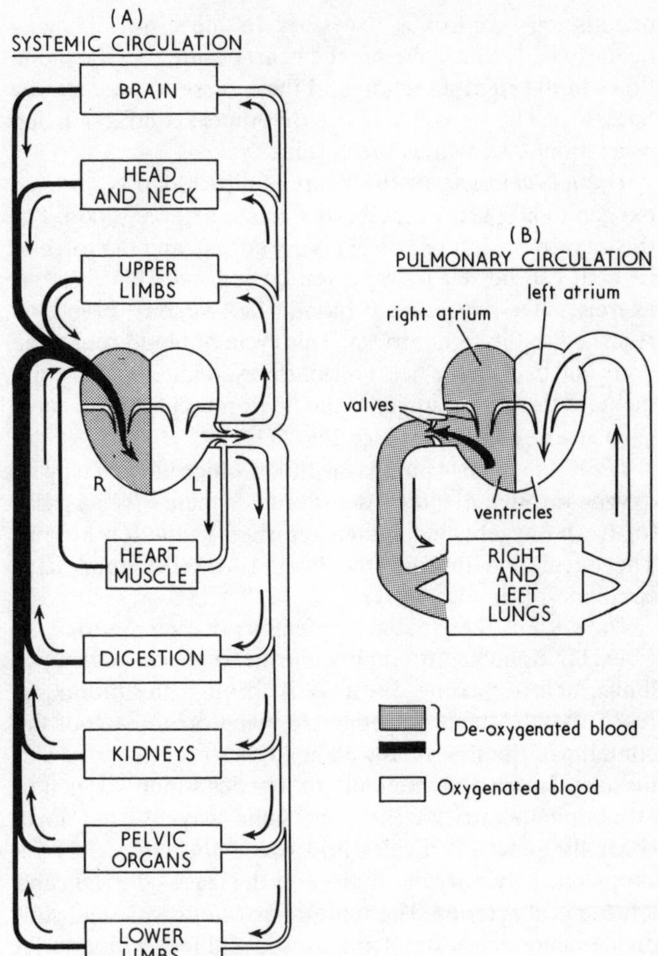

Fig. 5/1 A & B Diagrams showing A) the systemic circulation B) the pulmonary circulation

unconscious control is necessary to meet the changing needs of daily life. Valves in the heart ensure that the blood flows in the right direction and these close to prevent any backflow. The sequence of the rhythmical contraction and relaxation is known as the cardiac cycle.

The left ventricle of the heart pumps blood containing oxygen to all parts of the body – this is arterial blood. For this reason the left ventricle is very strong, and the force of its beat can be felt by a person during, or after vigorous exercise. De-oxygenated blood, i.e. venous blood, is returned to the right atrium. This cycle of blood round the body and back to the heart forms the systemic circulation; it includes the blood supply to the heart muscles by the coronary arteries and veins (*see* Fig. 5/1A).

The right ventricle pumps the blood which has given up its oxygen through the lungs in order to exchange waste gases for fresh oxygen; this is then returned to the left atrium. The circulation through the lungs forms the pulmonary circulation (*see* Fig. 5/1B).

The aorta is the great artery leaving the left ventricle. It gives off branches to supply the head, neck and upper limbs, before passing downwards through the thorax, to reach the abdomen. Arteries are given off to each of the contents of the thorax as well as to the thoracic joints and muscles. In the abdomen, major arteries supply oxygen to all abdominal structures in the same way. As the aorta nears the pelvis, it divides into two main arteries. These supply the pelvic organs, then enter the legs as the right and left femoral arteries. The femoral arteries divide and subdivide many times, supplying the leg and foot with arterial blood.

Arteries become smaller until eventually they are called arterioles which are very tiny vessels. These form a meshwork with minute capillaries, from which the veins are formed. This meshwork is called a capillary bed. Fluid plus oxygen, passes to the tissues from the arteriole end of the capillary bed, and is drawn back to the capillaries at the venous end (*see* p. 63).

Small veins unite to form bigger veins, eventually forming two very large veins, the superior vena cavae with blood from the neck and upper limbs and the inferior vena cava, with blood from the trunk and lower limbs. The venae cavae enter the right atrium. From the atrium, venous blood passes to the right ventricle. When the right ventricle contracts, venous blood is pumped into the pulmonary artery, and so through the pulmonary circulation.

Note: Arteries leave the heart and veins return to the heart, irrespective of the type of blood they carry.

Arteries

Arteries are tough; the walls are fibrous, with some muscular and elastic fibres. An artery remains open (patent) even if emptied of blood. The forward flow of arterial blood is ensured by the force of the ventricular contraction and the elasticity of the artery wall. No valves are necessary in an artery. The strength of the force decreases as arteries become smaller and become arterioles. This is due to the smallness of the many vessels and the distance from the initial force. The smallest arterioles are composed mostly of muscle fibres and they dilate or contract as blood to the area is required; this accommodation is controlled by the nervous system.

Veins

A vein differs in structure from an artery in that it is less tough; it is much softer and collapses if emptied of blood.

The formation of the two major veins has already been described. Deep veins accompany the arteries, more superficial veins do not always do so.

Venous return to the heart is started from the capillaries. These unite to form larger veins. The pressure in the veins is low, so movement of blood back to the heart is dependent on other factors.

Valves in the vein walls prevent backflow; these are needed especially in the veins of the lower leg. Changing pressures in the abdomen and in the chest during respiration exert a suction-like action on the veins, drawing the blood towards the heart; perhaps the most important factor is the active contraction of muscles. The muscles compress the soft walled veins lying between the muscle groups, propelling the blood onward. Prevention of varicose veins in the lower leg is chiefly dependent on muscular activity.

PHYSIOLOGY OF THE CIRCULATION

Blood is ejected into the systemic and pulmonary circulations at the same time and in the same amounts.

The sequence is as follows: both atria contract and after a short pause both ventricles contract, closing the valves between atrium and ventricle; the blood is ejected into the aorta and pulmonary artery. After a slightly longer pause the whole heart relaxes, the valve between the aorta and left ventricle closes and the pulmonary valve between the right ventricle and the pulmonary artery also closes. During this period of rest the heart refills to repeat the cycle.

Each contraction of the heart is initiated by a specialised node in the heart tissue, known as the pacemaker. In other words the heart beat is independent of a nerve supply from the central nervous system, while the rate, strength and amount of blood ejected are controlled by other factors which stimulate the cardiac centre in the brain. Normally the heart contracts about 70 times per minute, and discharges approximately 70ml of blood from each ventricle at each contraction. The flow in the arteries is in spurts – it pulsates; the flow in the veins is smooth and steady.

The pericardium is a double fold of membrane enclosing the heart, comparable with the pleura of the lungs. It enables the heart to beat with minimum friction.

Blood pressure

Every living tissue must be adequately supplied with arterial blood and hence a means of removal of waste products. The pressure in the large arteries is very high, but diminishes as the vessels become smaller and farther from the heart. Pressure is very low in the arterioles, capillaries and veins. To reach the brain, when the body is upright the blood has to flow upwards against the force of gravity and therefore the blood pressure in the brain is lower than at heart level. Very low blood pressure in the brain can result in fainting or loss of consciousness. In such a case the person should be placed flat; the head level with the heart. Brain tissue can only remain alive for a minute or two without its blood supply.

The pressure of the recoil of the elastic artery walls following the stretch imposed by ventricular contraction contributes to the blood pressure, the other factor is the resistance at the periphery, i.e. in the capillary bed area; this is known as peripheral resistance and plays an important part in the maintenance of the circulation. The peripheral resistance is under the control of the vasomotor centre in the brain.

The smaller vessels will open or close as need demands.

For example, during exercise more blood will be required by the active muscles, therefore vessels in the muscles will open while vessels in the digestive tract will close as temporarily they are not needed. Skin vessels will open as a means of heat regulation.

The pulse is the response of the arterial wall to pressure change as blood is ejected into the aorta; a pulse can be felt in any superficial artery. The radial artery at the wrist is the most convenient, and can be felt if gently compressed against the underlying bone. The pulse wave varies with the elasticity of the arterial wall. Pulse beat must not be confused with heart beat.

Arteries lie deep in the body and limbs to prevent damage or excess pressure during activity. At certain spots a pulse can be felt as the artery crosses a more superficial bony surface; these are the pressure points which are used medically.

Veins may be superficial or deep. The deep veins accompany arteries, the superficial veins are often near the surface and may be visible just below the skin.

Tissue fluid

The blood does not come into direct contact with the tissues, but through the medium of a fluid which bathes the tissue cells. Exchange of oxygen, food and waste (in solution) is by diffusion to and from the blood vessels via the tissue fluid. If tissue is damaged excess fluid may remain in the tissues, as oedema. Oedema may be caused through other reasons.

THE LYMPHATIC SYSTEM

Some of the fluid in the tissue spaces does not return directly to the blood stream. There are very small capillary-like vessels in the tissues running alongside the venous capillaries, and later accompanying the deep and superficial veins. These are called lymphatics (Fig. 5/2), or the lymphatic system. Lymphatics provide a channel for lymph which is different in composition from tissue fluid. Lymphatic vessels have valves and the lymph flows in the same direction as the venous blood. At intervals along the course of the lymphatics there are groups of nodes or glands. All lymph is filtered through at least one group of these glands before it is returned to the blood stream. Eventually, by joining together, two large lymph vessels are formed; these enter the venous circulation near the heart.

The lymphatic system is associated with the removal of bacteria or products of injury to the tissues, i.e. it is a scavenging system.

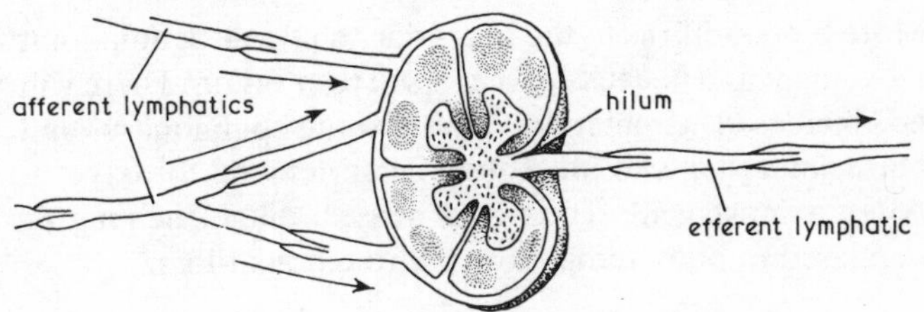

Fig. 5/2 Diagram of a lymphatic gland or node

Special cells capable of carrying out this function are manufactured in the nodes (glands), as well as in other specialised tissues. The filtering effect is evident when superficial glands, in the vicinity of an infection or injury, become enlarged and knotty, e.g. a septic finger may cause the glands in front of the elbow or in the axilla to become swollen and hard; in this case the course of the superficial vessel may be traced as a thin red line near the surface of the skin.

CHANGES IN CIRCULATION RELATED TO EXERCISE

ANTICIPATORY

The rate of the heart beat is increased, also the amount of blood leaving the heart; the pulse rate is increased and the peripheral resistance may rise at first. These changes are due to involuntary nervous control.

There will also be a redistribution of circulating blood with an increase to the heart muscle and skeletal muscles, and a decrease for digestion, excretion and the skin.

DURING EXERCISE

More blood returns to the heart due to muscle action, joint movement and because of increased respiration. There will be increased circulation through the pulmonary and muscular systems. to meet the higher demand for oxygen. After a short while cutaneous vessels dilate, helping to regulate the body temperature through heat loss.

AFTER EXERCISE

There is a gradual return to the normal.

TERMS ASSOCIATED WITH THE CIRCULATION

Haemoglobin A pigment in the blood which combines with the oxygen in the lungs, becoming oxy-haemoglobin. This is bright red in colour. After the oxygen has been released to the tissues, the haemoglobin is said to be reduced. This is much darker in colour, almost dark blue.

Anaemia A condition in which the haemoglobin concentration in the blood is below normal. Consequently insufficient oxygen reaches the tissues.

Cyanosis The name given to the blue tinge of lips, ears, etc. It is due to a high percentage of reduced haemoglobin in the blood and is seen in the superficial capillaries. There are many causes.

Varicosity or varicose veins A condition of inefficient valves in veins. They can be superficial or deep. A backflow of venous blood occurs, which pockets in the valve. The distorted veins arise most frequently in the calf of the leg.

Thrombus A clot of blood within a vein, usually adherent to the wall of the vein.

Thrombosis The condition of having a thrombus.

Blood pressure The lateral pressure exerted on blood vessel walls by the blood contained within them. The initial force comes from the ventricular contraction.

CHAPTER 6

The Digestive System

The digestive system is basically a tube-like structure through which food and fluids necessary for daily existence pass (*see* Fig. 6/1). During the passage down this tract, the food is prepared to be utilised or rejected by the body tissues. Each part of the digestive tract has a specific function.

THE MOUTH

The teeth and saliva prepare the food by breaking it down into particles, moistening and softening it, then forming it into a bolus before it is swallowed. A digestive juice in saliva begins the breakdown of starchy foods.

THE OESOPHAGUS

This is a tubular passage connecting the mouth and the stomach.

THE STOMACH

The stomach acts as a pouch-like receptacle, varying in shape and size. It lies beneath the diaphragm muscle and is protected by the lower ribs on the left side of the body. The lining of the stomach secretes gastric juice which continues the digestive processes. The muscular walls of the stomach contract and relax in a regular wave-like motion, to move the contents towards the exit of the stomach, and into the small intestine.

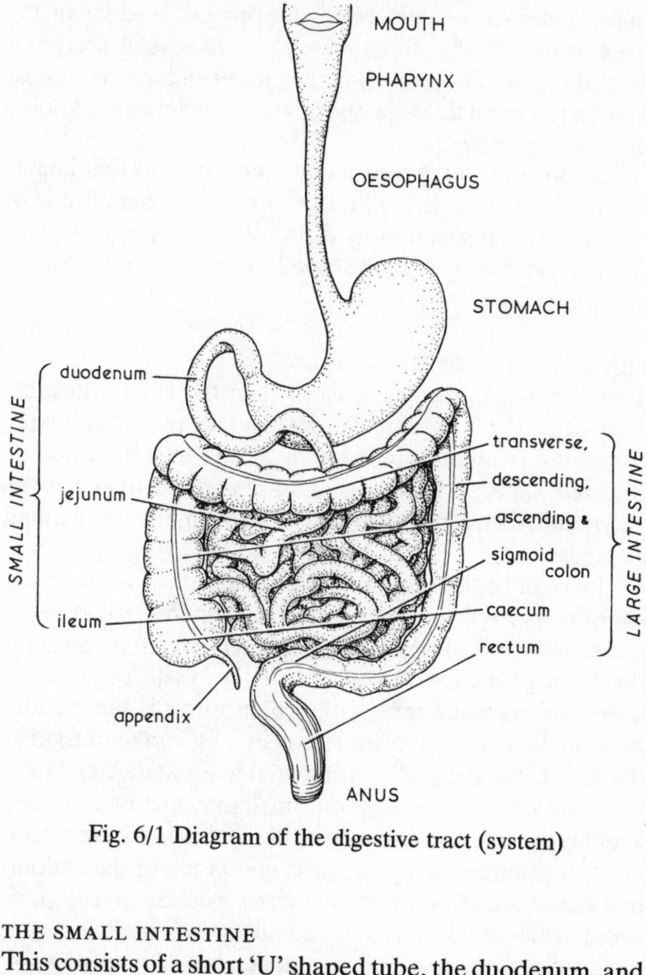

Fig. 6/1 Diagram of the digestive tract (system)

THE SMALL INTESTINE

This consists of a short 'U' shaped tube, the duodenum, and a much longer soft mobile tube, made up of the jejunum and the ileum. Externally there is no difference between the three parts. The small intestine secretes a digestive juice, and also receives bile from the gall bladder in the liver. Bile aids the digestion of fats. A special secretion from the pancreas joins the other juices in the small intestine. In the small intestine the final stages of the breakdown of food take place.

The mucous lining of the small intestine has fine finger-like projections, called villi; it is through the membrane of the villi that absorption of digested foods, fat and water occurs, and is taken into the blood stream for use or storage in various parts of the body.

THE LARGE INTESTINE OR COLON

This is much shorter and wider than the small intestine. Very little, if any, digestive processes take place here, but a considerable amount of water is absorbed. Unwanted or rejected material is passed along the colon to reach the pelvic end, the rectum. From here, waste matter is excreted as faeces.

Throughout the length of the small intestine there are continuous rhythmical wave-like movements called peristalsis; in the large intestine movement is irregular and takes the form of a mass movement of the waste products of digestion from one part of the colon towards the rectum. Mass movements are often stimulated by intake of food at the next meal. Peristalsis is essential to assist digested food from the stomach through the small intestine and thereby enables digestion to take place. Peristalsis is strong after a meal; it is improved by exercise, good tone of the abdominal muscles and indirectly by 'good' posture. Weak peristalsis leads to constipation and all that it entails.

The digestive processes are under involuntary control. Secretion of digestive juices is stimulated by anticipation, by an appetising smell, sight or even by the sound of cooking food; it may also be stimulated by the presence of food in part of the tract, e.g. food in the stomach will stimulate secretion in the small intestine; a 'starter' at a meal will stimulate secretion in the stomach. Mechanical movement is stimulated by bulk and by activity. Digestive processes are slowed down or temporarily stopped if the blood supply to digestion is required more urgently elsewhere in the body, e.g. if a child gets hurt and needs immediate attention.

Indigestion is liable to occur if vigorous exercise is taken too soon after a meal.

DIET

Diet must be considered in order to ensure adequate nutrition for the body tissues. The amount of food will vary with age, work or occupation, as well as physique, climate and fancy. Mental work requires a carefully balanced diet, the same as a diet for hard physical work, but the type of diet will be different. Essential requirements are:

Protein
Carbohydrates
Fat
Water
Vitamins and mineral salts

Protein provides energy for growth and the repair of tissue; proteins include meat, fish, eggs, cheese, milk, peas, etc.

Carbohydrates are the starchy foods and they provide energy fairly quickly. They break down into sugar for practical use; if not needed immediately they are stored as fat. They include bread, potatoes, sweets, etc., and are easily obtained. Carbohydrates are an inexpensive source of energy but they cannot replace protein.

Fat is either animal or vegetable and when digested, supplies energy or is stored as fat. A certain proportion of stored fat is necessary as insulation but too much is said to be harmful, causing a tendency to heart disease. Margarine from vegetable oils is said to be wholesome; other sources are butter, cream and fat meat.

Vitamins are supplied in most balanced diets. There are a few known vitamin deficiency diseases; e.g. lack of vitamin B may cause some forms of neuritis; lack of vitamin C may lead to bleeding into the tissues; lack of vitamin D may lead to a deficiency in bone repair.

Mineral salts do not liberate energy, but are essential for the normal functioning of various organs of the body. The need is small, and the loss constant through the skin and in the urine. A goitre, or enlarged thyroid gland in the neck, is sometimes due to lack of iodine in table salt. Lack of calcium can result in poor quality teeth or fragile bone structure, as well as loss of the 'clotting' property of blood.

Water is essential to maintain the body fluids. The body can survive for long periods without food, but not without fluid.

Roughage in the food is important. It provides the bulk already mentioned. Green vegetables and wholemeal bread provide a suitable source.

For those who wish to maintain a constant weight, food intake should balance energy output. Special diets are a specialised study and it is unwise to embark on anything more than minor adjustments without medical advice.

EXCRETION

Fluids containing waste salts in solution are excreted as sweat through the sweat glands in the skin. The kidneys are responsible for excretion of water, salts and some unwanted products of digestion which are in solution. Solids and some fluid are excreted as faeces. Moist gases are breathed out during respiration. It is important to realise the amount of fluid removed by one means or another, and therefore the need to replace this by ample fluid intake.

CHAPTER 7

The Nervous System

The nervous system is very complex. Without the nervous system no structure, not even a cell, in the human body could live or function. If a nerve is cut or diseased, the part of the body or limb supplied by that nerve becomes useless and wastes.

The nervous system is made up of a central part, and a peripheral part.

THE CENTRAL NERVOUS SYSTEM

The central nervous system consists of the brain and the spinal cord. Both structures have bony protection; the brain is enclosed in the skull and the cord in the bony canal made by the bodies and arches of the vertebrae, the vertebral canal.

The peripheral nervous system consists of the nerves passing from the spinal cord to the muscles; these are the motor nerves; and the nerves passing from the skin, joints and sensory organs back to the spinal cord; these are the sensory nerves. The peripheral nerves are more vulnerable than the central nervous system, as they are protected by soft tissue only.

The brain

The brain consists of two large masses of specialised nerve cells, the cerebral hemispheres or the cerebrum. These are enclosed in the vault of the skull. Voluntary movement is initiated from a special area of the cerebral hemisphere known as the motor area. The awareness of different sensations, e.g. pain or heat, or position of limbs, is registered in the sensory area. At the base of the brain there is a very important mass of nervous tissue, the cerebellum (*see* Fig. 7/1). This is concerned with balance and co-ordination.

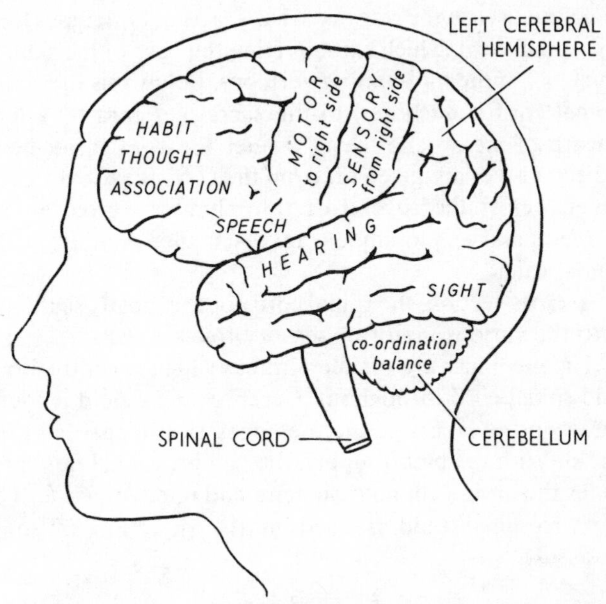

Fig. 7/1 Diagram to illustrate a cerebral hemisphere, the location of some of the special areas and the spinal cord

Special senses such as speech, hearing, memory and association are also represented in specific areas (*see* Fig. 7/1). Deep in the brain substance there are other special centres, already mentioned, e.g. the cardiac and the respiratory centre. The cranial nerves, e.g. optic for sight, auditory for hearing, are also partly contained within the brain tissue.

Formation of the spinal cord

Nerve fibres from the motor area join to form descending tracts; nerve fibres from the periphery form ascending tracts and reach the sensory area. Tracts of both types form the spinal cord, which extends from the base of the skull to the level of the mid-lumbar vertebra. Below this the actual spinal cord is anchored to the sacral vertebra by a long sheath of membrane which clothes the cord throughout. Paired nerves are given off from the cord at each vertebral level; below the second or third lumbar vertebrae the nerves pass very obliquely to reach their exit from the spinal canal.

In cross-section, the spinal cord appears neatly sectioned into the various motor or sensory tracts.

The meninges are the membranes which cover the brain and spinal cord throughout. Cerebro-spinal fluid is found between the layers of the meninges and in spaces in the brain. With the blood supply, the cerebro-spinal fluid provides the brain with nourishment, and removes waste. The cerebro-spinal fluid is used in the detection of some diseases.

Peripheral nerves

These are mixed nerves, i.e. motor and sensory nerve fibres

lie together within the same sheath; but the nerve impulses or messages pass in opposite directions. Peripheral nerves have mapped out courses. They usually run alongside an artery and vein before dividing into terminal branches to supply their objective. For the most part, nerves are deeply imbedded in muscle; in a few cases a nerve may become superficial across a joint, e.g. at the elbow to reach the forearm and hand; the sensory nerve causes a tingling in the fingers if the elbow is banged.

The motor nerves

Motor nerves arise from large grey cells in the motor area of the cerebral cortex (*see* Fig. 7/1). These motor fibres pass through the brain and form the descending motor tract in the spinal cord. At the level in the cord where the peripheral nerve is to be given off, a cell station is formed within the cord, i.e. the first nerve fibre ends and a second motor fibre makes a junction with the first. The second fibre passes from the cord, through the foramina between adjacent vertebrae, and becomes part of a mixed nerve. The peripheral nerve, as it is now called, passes to a muscle or muscle group, and supplies a nerve ending to each muscle fibre.

HOW A VOLUNTARY MUSCLE WORKS

Without a nerve a muscle cannot contract, so it is useless. Every muscle fibre will not contract for every movement (*see* 'all or none law', p. 18). Repetition and accurate practice ensures that only the essential muscle fibres work to achieve the required action and this results in economy of effort.

Sensory nerves

Sensory nerves are essential for a motor response. Sensations, such as touch, hearing or visual stimuli create the need or desire to respond by movement. Sensory nerve endings in muscle or joint, for example, send the information along the peripheral nerve to the cord, then via the ascending tract to the sensory area for interpretation; the motor nerves then cause the appropriate action to be taken (*see* Fig. 7/2). The whole action resembles a slot machine; the input is sensation, the output is action or purposeful stillness.

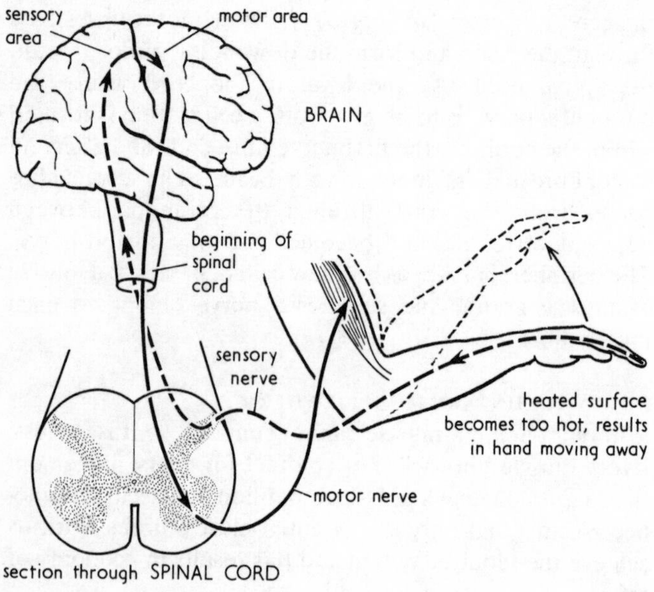

Fig. 7/2 Diagrammatic representation of a sensory stimulus and its motor response

It becomes obvious that sensory and motor nerves must function, and co-ordinate to produce purposeful movement. Loss of sensory nerves deprives the brain of the stimulus to 'do something'; loss of motor power prevents any response getting through to the muscles.

A reflex movement

A reflex is a protective mechanism. The shortest pathway between a stimulus and the response is taken subconsciously. Only after the movement has occurred does awareness reach consciousness; e.g. if one stands on a tin-tack the foot is withdrawn sharply, before pain is registered (*see* Fig. 7/3).

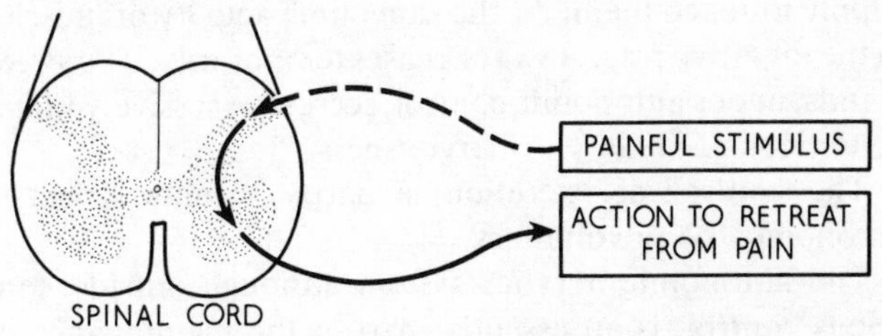

Fig. 7/3 Diagram of a section through the spinal cord showing the reflex pathway

THE AUTONOMIC NERVOUS SYSTEM

The autonomic nervous system could almost be called the 'automatic' nervous system as it functions below conscious level. It controls the activity of involuntary muscle in the heart, small blood vessels, secreting glands, e.g. the salivary glands, and in the skin. The autonomic nerve fibres arise in various parts of the brain and spinal cord, and ultimately pass to the organ, vessel or structure to be supplied.

Control of the activity of voluntary and involuntary muscle differs in the following way; a voluntary muscle contracts as the result of stimulation of the peripheral nerve to that muscle, and it ceases to contract when that stimulation is withdrawn; involuntary muscle has two antagonistic yet complementary nerve fibres, one to cause contraction, one for relaxation of the muscle. Therefore to preserve a 'middle' state these two fibres must be balanced in their activity. This is so during the normal pace of daily living. During exercise, it is the autonomic nervous system, acting subconsciously, which increases the rate of the heart beat, quickens and deepens breathing and dilates small blood vessels in working muscles to allow a fuller blood supply to reach them. At the same time activity of muscles in the digestive tract slows or ceases temporarily. The sweat glands, under autonomic control, secrete excessively under conditions of anxiety or nervousness.

The control of excretion is partly voluntary, partly autonomic, i.e. involuntary.

The autonomic nervous system although outside conscious control, is an essential part in the maintenance of healthy physical condition.

CHAPTER 8

Understanding Physical Disability

Disability does not necessarily go hand in hand with old age. It is true that on becoming elderly or reaching old age certain disabilities are inevitable although not always recognised. There are also many disabilities which affect the young and middle-aged, handicapping the individual in some aspects of daily living.

In this section the very briefest pathology is given for the purpose of enabling a leader to know the likely cause of any handicap. It is hoped to allay fears and give the leader confidence to accept a disabled person who would benefit from being a member of a normal class. It is important that whenever possible young disabled people should attend classes for their own age group, in spite of their handicap; they need to be stimulated to derive the maximum benefit and enjoyment from the class.

When or if it is necessary to form a class entirely of handicapped people, it is generally better for all concerned if those with different disabilities are grouped together. If a leader agrees to take such classes she should seek some advice and knowledge about the individuals, the content of the lesson and its presentation. With very few exceptions, disabled people know their limitations; they attend movement classes voluntarily for enjoyment, recreation and for the sense of well-being and fulfilment. There should be no attempt to make such sessions 'treatment' or 'therapeutic' in any sense, although the effect of their efforts will be beneficial.

CLASSIFICATION OF DISABILITY

Group 1 – Disability arising before or at birth or in early childhood.

This group are usually well-adjusted and adapted to their limitations and know their own capability.

Group 2 – Disabilities which tend to develop in middle age or later.

These conditions probably deteriorate as the person ages.

Group 3 – Disabilities associated with old age.

The mental reactions and responses of a disabled person, and certainly of the elderly, may be slow – but this does not imply that they are sub-normal. Their physical disability, partial sight, or poor hearing, may be the cause of their slow reaction.

Group 4 – Pathological causes

GROUP 1 – DISABILITY ARISING FROM BIRTH OR EARLY CHILDHOOD

To think that those in this group can be discounted is not quite true; but as they learn to cope with their particular problem, they probably enjoy being accepted without fuss and extra attention – this is as it should be. During the class, unobtrusive allowance should be made for inability to keep to a set tempo, in the use of hand apparatus, or in changing positions. The standard of effort should be maximal for each member of the class. The disabilities may include mild cerebral palsy, poliomyelitis, rheumatoid arthritis (Still's disease), respiratory problems or any of the congenital orthopaedic defects, partial sightedness or deafness.

GROUP 2 – DISABILITIES WHICH TEND TO DEVELOP IN MIDDLE AGE OR LATER

(a) Arthrosis

The middle-aged tend to show signs of 'wear and tear'. It is known as arthrosis – although most people talk of osteo-arthritis. It is a normal part of the ageing process and the symptoms vary from very mild to quite severe. Weight-bearing joints are chiefly affected, especially the hips, knees and lower spine; one joint only may be affected. Pain and stiffness of the joint first thing in the morning – also associated with cold and wet weather – are symptomatic. The muscles supporting arthritic joints tend to waste and lose some of their strength. Sufferers may complain of hips or knees 'giving way'. This involvement of joint and muscles sets up a vicious cycle (Fig. 8/1).

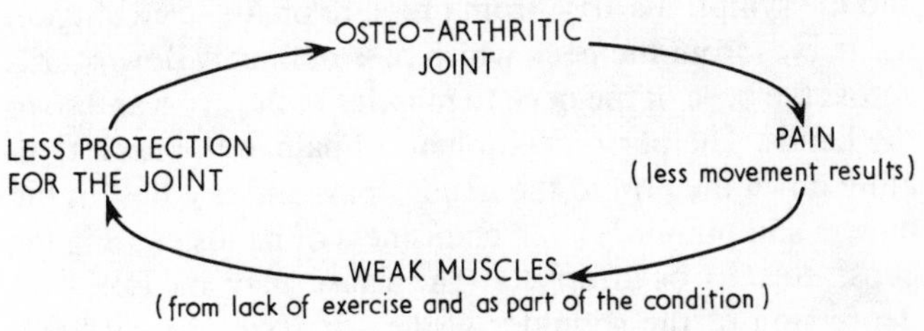

Fig. 8/1

Certain forms of exercise will help to relieve the symptoms. 'Once I get going I'm fine,' is often said; this is true. The muscles work, thereby maintaining some power to protect the joints – the circulation to the muscles and joint is improved, and the joint is prevented from becoming unduly stiff. Posture habits, or occupations involving much standing, are partly responsible for the condition of osteoarthritis. For this reason exercise or movement should be in sitting, lying or standing positions with stable support, i.e. off full weight. Standing still should be avoided, but walking and 'moving around' should be included to maintain independence and variety of activity.

N.B. Supports must be safe and steady, e.g. avoid a rickety chair. The support must also be of suitable height and easy to hold. Partner work is not advisable for those who need support; if one partner falls the whole class could topple like a pack of cards. Partner work for those who have no balance problems may be included but is of doubtful value as it tends to create tension.

(b) Cervical spondylosis: cervical osteoarthritis and conditions producing similar symptoms

These conditions are associated with middle-aged women and the symptoms arise from pressure on, or stretching of, the nerves from the neck which pass obliquely downwards across the base of the neck to muscles of the arms and so to the hands. The person complains of pain in the neck radiating down the arm to the hands, 'pins and needles' in the fingers and numbness and clumsiness of hands causing tea cups, etc., to be dropped. The cause may be excessive depression of the shoulder girdle – carrying heavy loads of shopping, unsuitable manual employment or heavy gardening.

A plastic or felt collar is frequently worn to minimise

movement of the neck, which irritates the nerves; collars should *not* be removed for exercise. A slightly raised position of the shoulder girdle often relieves the discomfort. Shoulder shrugging movements are good, provided the downward movement is gentle and never forced. Relaxation should also be encouraged – tension and anxiety are familiar characteristics of 'neck-ache'.

(c) Multiple sclerosis

It is in early middle age that the first signs of some neurological conditions may appear. With this particular condition the individual may complain of a leg feeling heavy or dragging unless a special effort is made, of fatigue out of proportion to the degree of activity, or of clumsy hands which is embarrassing to the person. There may also be emotional changes. The individual may laugh at very little – equally she may be upset or burst into tears for no reason. Fortunately neither situation lasts for long. If this occurs in a class little fuss is necessary; perhaps something to distract attention is sufficient.

The condition of multiple sclerosis is typified by alternate good and bad periods. At first many months may elapse between bad periods. Unfortunately the bad patches may become more frequent, balance becomes unstable, muscle weakness develops and sometimes the speech is slow or slurred. At this stage patience and understanding are very necessary. Although fatigue is complained about activity should be encouraged. While able to participate, these people enjoy group work and should be encouraged to be active; they will become less independent and may eventually attend in a wheelchair (*see* p. 104).

(d) Respiratory problems

In this age group respiratory problems are likely to be those continued from childhood, the bronchitis of old age not yet being evident (*see* p. 98).

GROUP 3 – DISABILITIES ASSOCIATED WITH OLD AGE

The Elderly: Chronological age has little significance, i.e. the most active individuals do not necessarily wear out first, or last – many factors contribute to 'ageing', e.g. interests, alertness of mind, friendship, will-power and heredity.

Like any other machine the human body deteriorates with age, but unlike mechanical machines the deterioration is not always related to a sedentary or an active life. Provided the over 80-year-olds take regular but sensible exercise and rest and maintain an interest in the world around them, they can and do enjoy movement sessions if suitably presented. It must be realised that there is sometimes a hereditary or family factor influencing the onset of joint, muscle and mental deterioration.

Many disabilities are not visibly obvious but the leader should be aware of their possible existence and of their nature. She will then be more able to prepare suitable work for a class of elderly people without causing discomfort to any of them.

The elderly may have any of the following disabilities although they may not always be obvious:

Raised blood pressure (arteriosclerosis)

This is probably due to loss of elasticity of the walls of the arteries.

Arterial circulation. The circulation or steady flow of blood through the body is maintained by pumping from the heart – through slightly elastic walled arteries. As the heart muscle relaxes and refills with blood ready for the next 'pump', the elastic artery wall recoils, maintaining a constant flow of blood forward. The more strenuous the activity the more blood is required by the muscle and the harder the heart has to pump (Fig. 8/2).

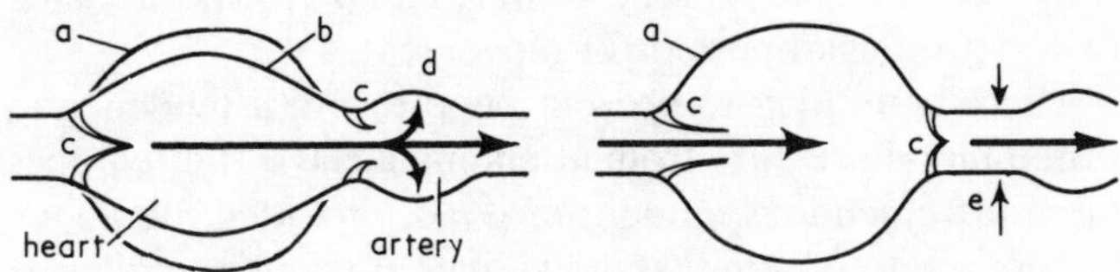

Fig. 8/2 Diagram showing how a steady arterial flow is maintained

a) = heart muscle relaxed b) = heart muscle pumping
c) = valves of the heart controlling direction of blood flow
d) = elastic artery wall stretched by force of pump
e) = recoiling artery wall ensuring continued blood flow

It is fairly easy to see that if the artery walls lose their elasticity and become hardened by age much extra strain is thrown upon the pump. Sometimes an adequate supply of blood to a part is not forthcoming – notably to head or arms as the flow has to travel upwards against the force of gravity. Because of this the action of changing from different levels (i.e. from sitting to lying and up to sitting or standing) can cause much distress. When such changes of position are necessary, they should be taken in each individual's own time. A lady has been known to stop attending a class rather than admit she could not lie on the floor without feeling ill. Many adults enjoy sitting on the floor, but lying flat is not recommended.

BASIC SYMPTOMS OF RAISED BLOOD PRESSURE AND MOVEMENTS LIABLE TO PRODUCE THEM:

Symptoms

Dizziness; faintness; faster and thumping heart beat; headache; sometimes affected vision.

Causes

Change of posture; work with the arms above the head or shoulder level; stooping; too short rest periods; too quick or too prolonged periods of movement.

There is nothing to prevent people with a tendency to raised blood pressure from attending a class. It often does them a tremendous amount of good, provided the above factors are noted and that individuals are taught to relax. If symptoms do occur, e.g. a dizzy spell, it should be suggested that a visit to the general practitioner (G.P.) would be advisable before the next class – just to make sure!

Venous circulation. Circulation of the blood through the body is completed by means of the venous circulation which conveys blood without oxygen back to the heart. The venous flow is brought about by the alternating pressures exerted on the veins by the movement of the muscles and joints, and other factors not so relevant to movement.

Loss of balance

This may be due to:

1 Deterioration of the balancing mechanism in the brain; poor vision; conditions affecting the inner ear; and stiff joints and weak muscles which are unable to adjust to change of position.

2 Difficulty of balance arises when standing with a narrow base, when transferring weight in any direction and

when movements involve stooping, especially if done quickly.

The arthroses (the more advanced condition to that discussed on p. 91)

Joints such as the hip joint or knee joint usually have glossy, friction-free surfaces, lubricated by a colourless fluid. In the wear and tear process the fluid tends to dry up, the once smooth joint surfaces become worn, and roughened areas appear so that the surfaces no longer glide on each other but grate; the range of movement is limited and often painful; and the joint capsule and other soft tissue lose elasticity. The affected joints become distorted or deformed and their function decreases.

Symptoms

1 Pain and stiffness of joints particularly after a period of rest or immobility.

2 Swelling and bony enlargement at the joint surfaces.

3 Wasting muscles resulting in less stable joints.

4 'Giving way' of joints while weight-bearing, leading to loss of balance and falls.

Main causes

Overweight predisposes to the condition; postural imbalance which is elsewhere in the body such as a short leg; slack posture; occupation – too much standing; injury or accident to a specific joint, or a hereditary tendency to osteoarthritis.

There will be days when these 'sufferers' are in pain or discomfort and will not be 'up to standard'. Encouragement such as small range movement and holding positions of the joint for short periods will be of benefit, although prolonged holding could lead to cramp. Often the relaxed atmosphere and general group activity soon enables the individual to overcome any disinclination to take part. Really painful joints are very liable to give way and therefore standing and moving around activities should be tactfully discouraged, or even not allowed until the painful stage has passed. Pain is a natural defence warning; it is a mistake to 'work off' pain. Stiffness without pain is another matter and is helped by movement of a pain-free nature.

Respiratory problems

Usually, 'breathing' is an involuntary movement unless something goes wrong with the mechanism, or an individual becomes temporarily 'puffed' due to extra activity. Respiratory problems can affect any age group but will probably relate more to the elderly. The necessary factors required for normal quiet breathing are:

Clear air passages (nose to lungs)

Sound lungs (air sacs and blood supply by which transfer of oxygen and waste gases can be effected)

A mobile chest which allows expansion of the lungs in all directions.

BRONCHITIS

This is the main cause of breathing distress in the elderly. The air passages which are normally open and lubricated by mucus, become irritated and inflamed. Excess mucus appears and is often coughed up. Frequent attacks of bronchitis lead to permanent changes in the bronchi (air passages) and they become less efficient. The bronchial walls become permanently thickened, narrowing the passage way (Fig. 8/3).

The breathing may become heavy and laboured in the effort to supply the necessary oxygen for the body tissues. Frequent pauses during activity may become necessary in order to regain breath. Shortage of oxygen affects all tissues of the body, the lung tissue itself and the heart muscle as well as the voluntary active muscles.

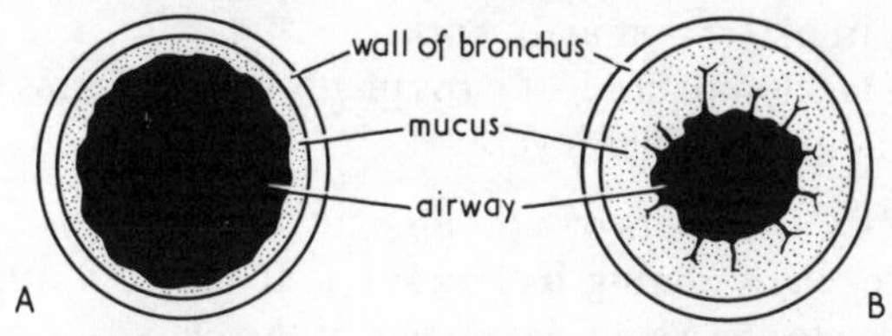

Fig. 8/3 Diagram to show section through a bronchial tube
a) clean airway b) bronchitic airway

Cardiac problems are frequently associated with severe bronchitis. The chest wall (thoracic cage) becomes less mobile and lung expansion is diminished. Anyone with respiratory problems is naturally anxious and tense and this perpetuates the symptoms. Anything to encourage or aid relaxation will benefit people with this distressing affliction. Relaxation may be general, or local for neck, shoulders and shoulder girdle.

N. B. As the rate and depth of breathing is very individual, any group, but especially an elderly group, should never be asked to do breathing exercises in a set rhythm or timing; the effect would be increased tension, holding of the breath and often quite real distress. The days of 'noisy' breathing are past; quiet, gentle, sighing expiration is preferred – inspiration being the natural sequence following the 'sigh out'.

Movements which normally increase the rate or depth of breathing to any appreciable extent should be taken very cautiously with all elderly people; these include:

Quick movements

Lengthy repetition of movements

Large movements of the hip, forward movements of the spine, and change of positions

Movements which might cause holding of the breath, e.g. strong trunk (abdominal) work

Too much repetition of movements with the arms above the head or shoulder level.

'Chesty' members of a group may be unable to attend a class regularly during bad weather. If they did attend it might undo the beneficial effects of the class.

ASTHMA

This is perhaps the most dramatic of all respiratory conditions. It can affect any age group, but usually appears in childhood or adolescence. The causes are many and varied. The sufferer usually has some warning of an impending attack. Asthma produces a characteristic tense posture, especially in the shoulder and neck region; the face shows signs of anxiety. The acute asthma attack is typical, but it should be remembered that mild breathing difficulties are always present. During an acute attack air can be taken in, but breathing out is difficult or impossible; the chest becomes more and more inflated and the sufferer more frightened for fear of losing their breath. If a leader suspects a class member has asthma it is advisable to enquire if any advice has been given to the person about procedure in case of an attack, or if therapy has been received at hospital; in which case the individual should be able to help her or himself. As little fuss as possible gives more confidence to the person concerned, as well as to the rest of the group. Relaxation is the most important asset in staving off or decreasing the severity of the attack. All easy rhythmical movements are very valuable. Movements increasing breathlessness or tension should be avoided. Cold and damp weather affects people with this disability. One great benefit derived from attendance at movement sessions is the confidence gained and the improvement in posture due to relaxation.

EMPHYSEMA

(This should not be confused with empyema which is a septic condition of the chest.)

Emphysema is fairly common as an elderly person's complaint and it is also very variable in its degree of severity. Breathlessness on exertion is the main problem – very similar to the bronchitic but without the cough. The tissues of the air sacs lose some elasticity; they become stretched and merge into fewer and larger sacs (Fig. 8/4).

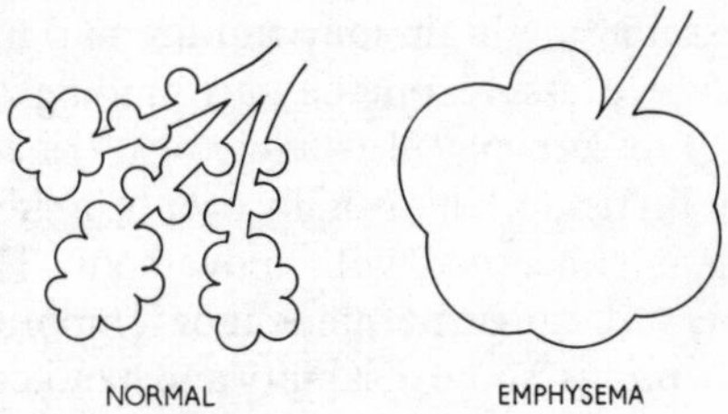

Fig. 8/4 Diagram of normal and abnormal air sac

It can be seen in the diagram that the surface to which the air is exposed is diminished, resulting in less oxygen exchange; because of the relative inelasticity of the air sacs, stale air is left in the lungs. The symptoms may be minimal and put down to old age, but the condition does become more pronounced. Those with very severe symptoms will not attend classes; the moderates must go their own pace and they should benefit from any activity which does not cause them to become breathless or distressed. There may be associated strain of the heart muscles.

As a general rule all people with respiratory problems will benefit from gentle easy movement, producing relaxation of the muscles and reducing nervous tension. Controlled breathing helps to maintain the mobility of the chest and clears the air passages and lung tissue.

GROUP 4 – PATHOLOGICAL DISORDERS

No attempt is being made to describe the pathology of the following conditions, only the basic reasons for the resulting disabilities and some of the neurological disorders.

Hemiplegia (stroke)

A blood vessel in the brain may rupture and the resulting bleeding cause pressure on the surrounding area of the brain tissue. Pressure may also be the result of a tumour, or disease of brain tissue. This usually occurs in the part of the brain which is concerned with movement. The damage from this pressure can be permanent or temporary but with very few exceptions, some disability and handicap remains. The characteristic feature of a 'stroke' is a one-sided paralysis or weakness, involving both limbs but in varying degrees. Facial muscles and speech may also be affected. Pain and abnormal sensation may be present.

Recovery from the initial catastrophe occurs slowly. The individual can be seen to improve for several months after the initial attack; the leg usually improves more than the arm. No order of timing of return to power can be made, but recognition of effort and encouragement are of infinite value and importance. The mental reaction appears to be slow, because the nerve paths between the brain and the limb are damaged; the physical response is slow, causing much frustration.

People who have had a stroke are *not* stupid and they should be talked to and treated as rational human beings; never talk down to them but obviously make allowances for their handicap. Sometimes the leader can be too kind or sympathetic and not enough is expected from the individual – they should be admired not pitied.

Those with speech disability know what they want to say, and suffer the upsetting experience of hearing the wrong words come out.

The condition itself limits the possibility of all movements and there is no danger of doing too much; indeed the majority should be encouraged to do more, though there are always exceptions. Balance will be difficult in sitting or in standing, due to defective muscle power and sensation on the one side. The use of both arms should be encouraged but the individual must not be worried. Help may be needed to take off a jacket, etc.; clothing can be an added restriction to movement, and jackets, pullovers, etc., should be removed when the temperature of the room or weather permits. Walking aids, i.e. sticks or tripods, may be used and should be kept beside their owners, not left with hat and coat. People who have had a stroke should not be allowed to clutch at furniture or, worse still, another member of the class. Great patience is necessary when helping a hemiplegic person; they can hold up the class but nevertheless must not be hurried or ignored since they tend to get very flustered.

Parkinson's disease

Some specialised small areas in the brain are affected in this distressing condition. The main signs which may make a person appear to be different are:

1 The absence of the automatic swing of the arms in walking
2 The body appears rigid; walking is stilted and tends to be in a straight line, getting faster; the sequence is – 'walk, stop abruptly – turn sharply – stop – walk' – this gives walking a mechanical quality
3 All movements are jerky or sometimes inhibited
4 The facial muscles are affected; wrinkles are absent, the eyes do not blink and natural emotional changes are not registered, i.e. no smile, no sign of recognition or of understanding. This is perhaps the most distressing symptom for the sufferers as they are often considered dull, uninterested or stupid.

All the above are due to muscle stiffness.

Usually these people are intelligent and easily hurt by being misunderstood. It is very good for them to be in a normal class and to be treated as such. The leader must make allowances for slow reactions which are physical not mental. Speech may also be slurred and slow. Medical or surgical treatment for 'Parkinson's' is of value for some people, relieving certain of the symptoms.

Multiple sclerosis

This has been introduced already in Group 2. Now the disease may either not have progressed very much, or the individual may be living a chairbound life; either way, it is most important for these people to do as much as possible.

Every encouragement is needed since many do as little as they can get away with and leave making much effort until it is too late because movement may be so difficult. Muscle weakness leading to complete inability to perform certain movements may overtake them; either arms *or* legs may be most affected. With this condition goes an apparently light-hearted attitude and a non-realisation of the deteriorating situation, even though the present handicaps are a worry. Inwardly they do not feel 'light-hearted'. Real fatigue should be avoided. All exercises should be attempted in the wheelchair. Alternate hip lifting, using the waistline muscles (rocking), teaches taking the weight off the 'seat' bones, which tend to get sore sitting all day. If the balance in standing is unsteady, movements in this position should be discouraged allowing the person to enjoy a really stimulating session in their chair. If aids are used they must be to hand. In the later stages vocal communication may become quite a problem. Incontinence is often present and must be treated very sympathetically.

Muscular disorders

A few class members may have marked muscle weakness and wasting but do not appear to have any other features of a neurological condition: the probable cause is a specific dysfunction of the actual muscles, the nervous system being normal. Affected muscles cannot contract, they lose their bulk and are unable to move joints or support the trunk; these people appreciate a good session in wheelchairs. Fatigue must be avoided.

Paraplegia

The causes may be neurological or orthopaedic. If orthopaedic, i.e. fracture of the spine, the result will be paralysis of both legs and the trunk below the waist. The condition does not necessarily get worse. For these, life in a wheelchair can be very strenuous and exciting. Usually the upper trunk and arms are very strong. Everything in a normal class should be attempted. If splints or calipers for the legs are worn, it is not really the province of the leader to assist those who are unable to walk independently. Incontinence may be a problem and an appliance may be worn. Sensation may be abnormal or absent.

Orthopaedic disorders leading to disability

Orthopaedic problems are structural i.e. bony or joint conditions or those 'wear and tear' conditions which have warranted surgery, e.g. gross osteoarthrosis of the hip joint. Only good can come by movement provided it is pain-free.

Painful back conditions may be osteoarthritic in nature and can be considered as before (see osteoarthritis p. 91). It is very difficult for sufferers to become 'off weight' for movement (as for hips or knees); even in sitting the weight of the body must fall through the spinal joints (*see* Fig. 8/5).

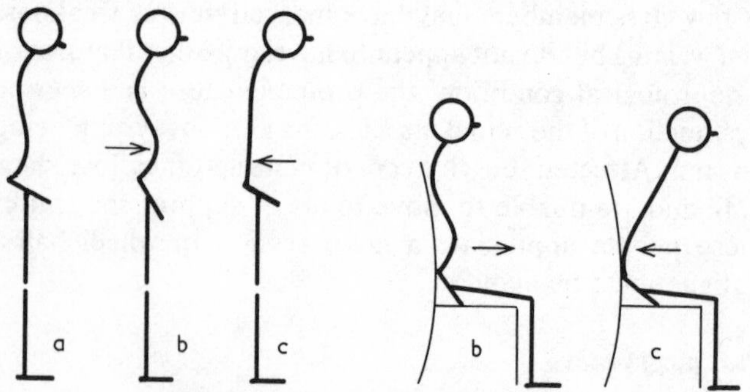

Fig. 8/5 Diagram of pelvic rocking
a) normal spinal curve b) hollow back c) flat back

Gentle tipping of the pelvis from the 'waist' in sitting or standing helps to maintain mobility of the lower spine and to delay postural backache. The movement is achieved by alternate contraction of the abdominal and back muscles (Fig. 8/5).

Disc lesions are rarely incurred in the elderly but may be residual from earlier days. Any back sufferer will probably, though not necessarily, wear a corset to support the spine. The purpose being to limit or prevent movement of the painful area and to give support when muscles can no longer provide it. Supports should not be removed for class work – it is surprising the amount and range of movement possible for other parts of the body while wearing the support. Attempts should be made to use the trunk and postural muscles as much as possible.

Painful feet can cause hobbling and poor balance. Supporting shoes provide comfort and help. 'Heels' or 'no heels' must be an individual decision; usually a reasonably low broad heel is safer and more comfortable. As people grow older the tendency is to turn the feet outward. As far as safe balance and comfort can be maintained, they should be encouraged to point the feet forward (in the direction they are going), so that when walking the propulsion forward from the back foot is in the line of the big toe joint – not off the side (*see* Fig. 8/6).

Very few leaders think of including *toe movements* in their lesson. To incorporate these movements may be difficult in practice but they are important. Toe movement can be made within shoes but when possible the shoes should be removed. Rigid toe joints are associated with weak muscles in the foot which lead to painful feet, poor balance and uncomfortable walking.

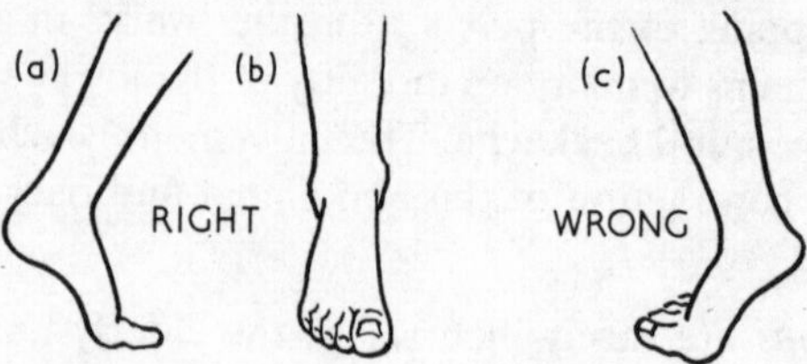

Fig. 8/6 Diagram to show detail of 'push off' through direct line of joint of big toe
a) and b) correct c) wrong

N. B. Try standing on one leg watching how the toes have to grip, adjust to balance and to the floor surface, and imagine the difficulty if unable to use the toes.

The signs and symptoms of any orthopaedic condition become more evident as age advances and muscle tone and power diminish.

Chiropody is available at clinics for those with nail or callous problems.

Glossary

Arthroses

(a) Osteoarthritis is a bony joint problem usually concerning the bigger joints. Movement helps osteoarthritis.
(b) Rheumatoid arthritis is a joint problem which affects mainly the soft tissues surrounding the joint, i.e. ligaments and capsules. The bony formation of the joints are also soon affected, becoming distorted; usually the smaller joints are involved, i.e. fingers, wrists or toes. Movement is not suitable when pain exists. Splints are sometimes worn to delay deformity.

The form affecting children is known as Still's disease and usually leaves severe handicaps.

Cerebral Palsy

This is damage to part of the brain before or at birth. There is paralysis (palsy) of one or both sides of the body – sometimes only one limb or part of a limb is affected, e.g. lower leg only. The type of paralysis is spastic – hence the adopted name for these children – spastics.

Orthopaedic

This is the term relating to bones, joints and associated soft tissues such as ligaments.

Paralysis

This may be motor or sensory; either nerve may be impaired or lost resulting in inability to use a muscle group, or to appreciate sensation in an area, e.g. the person who cannot feel heat and often suffers burns; he may not be able to register pain, and may stand on a nail and get a septic sore.

Paraplegia

This is paralysis or weakness of muscles below the waist due to damage to the spinal cord through sports injuries, or accidents involving spinal joints; the causes may be orthopaedic or neurological as in cerebral palsy, spina bifida or the later stages of multiple sclerosis. Abnormal sensation and incontinence are major complications.

Paresis

This is a weakness of muscle due to nerve injury.

Pathological

This is a term which includes the signs and symptoms due to disease processes which may sometimes be the result of an infection, as poliomyelitis, or to unknown causes, as in multiple sclerosis.

Poliomyelitis (old term – **infantile paralysis)**

This is an acute infection attacking a specific area of the spinal cord and resulting in flaccid paralysis of groups of muscles in no set pattern. It usually affects children and young adults but is rarely seen in the acute stage in older people. Leaders may well find people in their classes who contracted it many years ago and who have resultant disability. The residual paralysis is irreversible. Flaccid is floppy – the opposite of spastic.

Spastic

This is excess tension of muscles and stiffness of joints due to a neurological cause, e.g. cerebral palsy or hemiplegia.

Spina Bifida

This is a developmental defect of the lower spine, in which the bony canal containing the spinal cord is not complete, leaving an area of cord and its coverings exposed. The resulting damage to the cord may vary greatly, and may be a cause of paraplegia. Very mild damage may result only in weakness of leg muscles. Spasticity is often a feature, and incontinence may be another problem. Sensation in affected parts is absent or abnormal.

Weight-bearing

This refers to joints through which the body weight is taken, i.e. in standing, the weight-bearing joints are: feet, ankles, knees, hips, sacro-iliac joints, and all the joints of the spine; in sitting, the sacro-iliac joints and the joints of the spine only are involved. In lying the body weight does not fall through any joint.

Index